Lean Six Sigma and the Theory of Constraints

This book serves as a valuable resource for understanding the powerful synergy between two essential improvement management methodologies—Lean Six Sigma (LSS) and the Theory of Constraints (ToC). The authors expertly illustrate how integrating these approaches can result in substantial enhancements in organizational efficiency, particularly within the realms of continuous process improvement and key operational performance indicators. This comprehensive book offers a deep dive into the potential benefits of merging LSS and ToC, providing readers with practical insights and strategies for achieving optimal results in their business operations.

Lean Six Sigma and the Theory of Constraints: Tools, Case Studies, and Synergies uniquely combines two powerful methodologies, LSS and ToC, demonstrating how their integration can enhance operational efficiency and drive superior business outcomes. By providing a wealth of research findings, this resource substantiates the theories and methodologies with real empirical evidence. It establishes a robust theoretical framework for both LSS and the ToC, equipping readers with a comprehensive understanding of these methodologies to effectively implement them in practical settings. This book includes case studies and essential tools such as the DMAIC cycle, VSM, Kanban, Five-Step Focus, and bottleneck identification tool, among others.

Readers will gain insights on how to leverage these tools to identify waste, reduce variation, and eliminate constraints, resulting in tangible enhancements within organizations.

Lean Six Sigma and the Theory of Constraints

Tools, Case Studies, and Synergies

Wieslaw Urban and Patrycja Rogowska

Routledge
Taylor & Francis Group
NEW YORK AND LONDON

Designed cover image: Shutterstock

First published 2027
by Routledge
605 Third Avenue, New York, NY 10158

and by Routledge
4 Park Square, Milton Park, Abingdon, Oxon OX14 4RN

Routledge is an imprint of the Taylor & Francis Group, an informa business

British Library Cataloguing-in-Publication Data
A catalogue record for this book is available from the British Library

ISBN: 978-1-041-10410-0 (hbk)
ISBN: 978-1-041-10474-2 (pbk)
ISBN: 978-1-003-65523-7 (ebk)

DOI: 10.1201/9781003655237

Typeset in Palatino
by codeMantra

Contents

Foreword

Ten years ago, I wondered what kind of organization would emerge if it had completed the so-called lean transformation, because the literature wasn't clear on the matter. We used a very broad definition of lean management, which is why we called such an organization a continuously improving organization. This is an organization in which, with an infinite, systematic, and cyclical approach, increased value is added, and waste is eliminated in striving for perfection. Everyone is involved in improving products, services, or processes across all systems. The result is sustainable improvements in effectiveness and efficiency, for example, quality, speed, flexibility, costs, and sustainability.

In our literature review, we examined Lean Management, Six Sigma, and Theory of Constraints, among other approaches. These approaches have different origins and therefore employ different principles and tools. Nevertheless, we found that such organizations share several common characteristics. For example, the process is central and there is a strong focus on the customer. Improvement is approached in a structured manner, and employees are involved in the improvement process.

I've worked extensively in food companies on improvements using Lean, Six Sigma, and Theory of Constraints. There, I saw that improving is initially difficult and occurs ad hoc. If the organization persists in improving, things will get better over time. However, new improvement methodologies are often embraced. This year, lean is the focus, followed by Six Sigma, and then Theory of Constraints. This may be justified, for example, because environmental conditions have changed, but even then, it leads to improvement initiatives that are left unfinished and employees who become frustrated and cynical wonder what training course management followed this time. In such cases, it helps to have a framework that combines different methodologies and in which the various tools reinforce each other. For example, Kanban and drum-buffer-rope can reinforce each other, and value stream maps can also reveal bottlenecks.

I introduced lean management a long time ago in the mecca of Six Sigma at the time: General Electric, and I know from experience that combining methodologies isn't always easy. It does, however, enrich the operational options and ensure a conscious and balanced application of principles and tools. Currently, there's growing interest in connecting the available methodologies and applying them in an integrated manner. Therefore, this book is needed right now.

Structured, with the right concepts and good practical examples, Wieslaw Urban and Patrycja Rogowska guide the reader through Lean Management, Six Sigma, and Theory of Constraints, integrating the three methodologies into a cohesive framework that is elaborated on with practical examples. With this framework, organizations can begin improving where the challenges lie but not stop there. All surrounding processes must also become predictable and reliable, so that true value is created and waste is eliminated.

Dr. Ton van Kollenburg
Professor of Impactful Value Chains
Avans University of Applied Science, The Netherlands

Preface

In an environment marked by growing complexity, cost pressures, and elevated customer expectations, alongside the emergence of strong competitors from new regions, adopting a continuous improvement approach is essential for building a sustainable competitive advantage. Methods and tools that not only identify sources of inefficiency but also effectively and sustainably eliminate them across the organization are becoming increasingly vital.

Companies that recognize the importance of systematically improving their processes are increasingly adopting proven management approaches such as Lean Management, Six Sigma, Lean Six Sigma, and the Theory of Constraints. While each of these methods delivers measurable results when applied individually, there is growing interest in their simultaneous, integrated application.

Lean Management, Six Sigma, Lean Six Sigma, and the Theory of Constraints are among the most well-known and widely used process improvement methodologies worldwide. Despite their popularity, companies often face significant challenges when implementing these approaches, and the anticipated results are not always fully realized. There are growing concerns about how to integrate individual approaches within a single organization, how to achieve synergistic effects while preventing the mutual weakening of these approaches, and how to avoid chaos by establishing an efficient and rapidly adaptable organizational system.

This book was written not only to organize and consolidate knowledge from these fields but, above all, to present solutions and original concepts that foster synergy among these methodologies. Readers will become familiar with the fundamental principles of Lean Management, Six Sigma, and the Theory of Constraints. However, the authors focus primarily on the most critical issues related to their practical application, with particular emphasis on their integrated implementation within an organization. They propose an integration framework for the three methodologies and demonstrate how their combination can produce mutually reinforcing effects. In this context, the concept of Constraint-Driven Lean Six Sigma is developed.

A distinctive feature of this book is its grounding of key thoughts in empirical evidence. According to the authors, expert knowledge is fully valid only when it is based not only on in-depth insight into the functioning of companies but also on robust empirical data that supports the arguments. The clearly presented research demonstrates, among other things, the critical importance of simultaneously employing multiple approaches.

This book explores a curated selection of methods and tools drawn from a broader range, tailored to facilitate the implementation of a synergy-based approach, accompanied by practical tips. It also introduces original solutions, including a multi-method model for identifying bottlenecks within organizations, enabling effective resolution of the often-challenging task of accurately pinpointing system constraints. The thoroughly explained methods and tools are further reinforced by case studies that illustrate the practical application of elaborated concepts.

This book offers readers structured theoretical knowledge alongside practical guidance to enhance their understanding of the integrated application of process improvement methods. It is important to note that combining these methods requires careful planning and adaptation to the organization's specific characteristics to fully harness their synergistic potential.

Authors

Wieslaw Urban works at the Faculty of Engineering Management, Bialystok University of Technology, Bialystok, Poland. He is the Head of the Production Management Chair. His research focuses on several issues related to operational business development, namely, Production Management, Lean Manufacturing, Quality Management, and similar areas. He has published nearly 200 scientific papers and several books. His studies are published in journals ranked by Journal Citation Reports, and his publications are cited internationally. His Google Scholar H-index is 20. He has supervised six doctoral dissertations. He often serves as a reviewer for scientific journals.

Professor Urban also serves on several scientific boards and provides expertise for several domestic and international institutions, namely, the European Commission, NCBiR (Polish public agency for applied research), and PARP (Polish public agency financing innovations). He intensively provides training and consultancy for organizations, including widely known domestic and international corporations. He was the project coordinator of the GoSmart BSR project, financed by the Interreg BSR program on S3 development and business internationalization, with a budget of €1.8 million, 2018–2021. He was also the project coordinator of another Interreg BSR project, GoSmart & Excel BSR, with a budget of €0.8 million, 2021–2022.

Patrycja Rogowska is a researcher at the Faculty of Engineering Management, Bialystok University of Technology, Bialystok, Poland, where she is associated with the Production Management Chair. Her research interests include production management, theory of constraints, and process improvement in manufacturing systems. She is the author of numerous scientific publications and co-author of the book *Value Stream Management: A Source of Sustainable Benefits in an Organization*. Her scientific achievements are widely cited in the literature on the subject.

In 2023, she defended her doctoral dissertation in the field of management and quality science. Her doctoral thesis, titled "Methodology for Identifying Constraints in Production Systems in the Printing Industry," focused on developing a methodology for diagnosing bottlenecks in complex production environments.

Dr. Rogowska was also a member of the GoSmart BSR project team, funded under the Interreg BSR program. The project, implemented between 2018 and 2021 with a budget of €1.8 million, focused on the development of smart specializations (S3) and the internationalization of business in the Baltic Sea region.

Acknowledgments

We would like to express our sincere gratitude to our colleagues at the Chair of Production Management, Faculty of Engineering Management, Bialystok University of Technology, for their support and for fostering an open academic environment that facilitated the exchange of ideas and the development of this book.

We would like to thank Professor Ton van Kollenburg for his foreword and inspiration.

We would also like to express our sincere gratitude for the mutual help, inspiration, perseverance, and collaborative effort put into preparing this book.

We also acknowledge that the field research presented in this book was conducted as part of the team research project WZ/WIZ-INZ/4/2025 at Bialystok University of Technology.

1

Introduction

1.1 Objectives of This Book

Effective operations management, combined with the systematic improvement of organizational processes and their targeting toward particular business effects, is fundamental to winning the competitive game and ensuring business growth. Organizations that recognize the indispensable need for process improvement often turn to Lean Management or other effective methodologies. While process improvement based on a single methodology can be effective, there is a growing interest in leveraging multiple approaches simultaneously.

Main process improvement methodologies, namely, Lean, Six Sigma, and the Theory of Constraints, are generally well known. These methodologies are widely practiced by organizations all over the world. Many experiences have been gathered by various types of organizations, and plenty of research has been conducted by academics. However, there are still unmet expectations and numerous difficulties in using them. A special challenge is the combined use of the above methodologies.

This book has several meaningful goals. First, it aims to provide a comprehensive explanation of the peculiarities of Lean Management, Six Sigma, Lean Six Sigma, and the Theory of Constraints. However, the purpose is not to review all the plentiful management approaches but to discuss only the essential and specific issues, particularly those crucial when using them together within an organization. Next, this book aims at building a framework for the coexistence of these three methodologies, highlighting how they can strengthen each other. A concept of Constraint-Driven Lean Six Sigma is proposed.

The final aim, no less important, is to operationalize the joint functioning of methodologies within an organizational system at the tool level. Several important methods related to this objective will be explained, along with hits on how to adopt them. Where necessary, original frameworks are proposed, such as a multi-method framework for bottleneck identification; it refers to multiple techniques, which are provided with detailed insights and

DOI: 10.1201/9781003655237-1

clarifications. This book addresses practical issues by presenting empirical case studies alongside business case studies.

Throughout all the chapters, the authors try to demonstrate effective process improvement based on the common use of three methodologies. The combination of the three is not automatic but requires reasonable preparation to maximize benefits, seeking synergies along with fitness to the individual development path desired by an organization.

1.2 The Three Methodologies Working Together

Each of the methodologies has its own story and narratives, along with a peculiar manner of describing organization and its business reality. But first of all, each of them has different origins. However, over the years of operation, they have managed to borrow a lot from each other; hence, there are also many common elements. Among many reports, opinions, and slogans, one can miss what is fundamental in these methodologies.

As indispensable for further parts, Chapter 2 provides a comparison of the peculiarities of Lean, Six Sigma, and the Theory of Constraints. An analysis of the origins contributes to understanding how these methodologies work today and what kind of so-called management philosophies they imply. Particular attention is devoted to the elaboration of fundamental principles for each approach. As in the case of Lean Management, the literature tries to extract management principles for better explanation; for the other two methodologies, principles are derived in a similar way.

Managers desire effective methods that allow their processes to work better and better. Searching for the key factor that makes process improvement yield high results is therefore of great importance. Chapter 2 also presents an investigation of a large sample of medium and huge companies, according to which employee engagement is a factor distinguishing between good and poor performance in process improvement. Regardless of the methodologies employed, an organization must strive to achieve employee engagement if its efforts to improve processes are not to be undermined.

The elementary framework of the joint operation of Lean, Six Sigma, and the Theory of Constraints can be developed by starting from the organizational systems typical for each of them. How each methodology practically works in everyday organizational reality is the issue addressed in Chapter 3. Six Sigma is the most structured among them when it comes to roles and responsibilities defined in the organization. The challenges and shortcomings of the methodologies are analyzed as direct motivators for organizations to use multiple approaches.

The investigation of medium and large companies clearly shows that they often employ more than one improvement methodology. The Theory of

Constraints appeared as the most commonly co-used approach. A structured framework of joint functioning of the methodologies would be beneficial. The Constraint-Driven Lean Six Sigma concept highlights the elements that are crucial for the joint use of the three methodologies. A smart combination of these three systems in an appropriate way, with the right emphasis and use of appropriate elements, can lead to a highly effective system for improving processes in the organization.

1.3 Value Created and Value Lost

As the notion of value is capacious, it gives real meaning and efficacy to process improvement efforts. Value is interwoven in all activities within the organization and its contacts with external partners. The value in the eyes of customers is always the starting point for defining what is value and what is not in the organization. If improvement actions and projects are to bring significant results, value has to be considered the foreground navigating concept. Process improvement methodologies take the concept of value seriously and offer a framework for uncovering and understanding it, along with operational tools supporting it.

The concept of value with reference to process improvement is explained in Chapter 4 of this book. Along with value, hand in hand, goes the concept of muda (waste). Muda presented by Toyota's guru Ohno is still valid and inspiring; however, it might be perceived as instructional catchphrases. The chapter argues that these "teaching slogans" should start meaning something in company's everyday life—if the improvement would go to reality.

The concept of constraints is well harmonized with value, as it generally aims to avoid wasting value in the processes along the path of value creation. The issues of what is a constraint along with what is a bottleneck are clarified. Several methods for bottleneck identification are described. Presented business case study deepens the understanding of how to determine where the bottleneck is in the organizational system.

Further explanations of constraints, the main issue in the Theory of Constraints, are the subject of interest in Chapter 5. Many possible constraints of the organizational system are discussed. Another empirical investigation on a sample of printing companies provides interesting insights on real constraints limiting the growth of companies in an industry. The field data made it possible to organize the constraints into some specific types. Structuring constraints is intended to provide managers with a broad perspective on the constraints they seek in their organizations.

The issue of "what" value is has been explained in the previous chapter, but Chapter 6 provides an explanation of how to find "where" value is produced, along with where it is wasted by the organization. The value stream is a well-suited concept for recognizing value loss and value creation points, conditions, and peculiarities. Value stream mapping, a technique for determining value and designing improvements, is comprehensively explained, provided business case study enables a better understanding of this issue.

As a bottleneck of a stream may seem as obvious, however, in reality of company's operations its identification is not so easy. The multimethod framework for bottleneck identification is proposed. This practical guide helps in determining the real point performing as a bottleneck of a stream. This chapter provides an example on how a bottleneck is identified by following the framework.

1.4 Elevating the Process Efficiency Step by Step

The final two chapters, Chapters 7 and 8, are devoted to the main organizational routines related to introducing improvements in the organization, as well as to introducing and maintaining steady streams within the organization. These chapters are intended to supplement the previous chapters and, in particular, to propose further solutions integrating the three approaches at the level of organizational routines related to improvements.

Each of the methodologies is a proven procedure for dealing with improvements. Even though the general goal is very similar for all of them, each differs significantly in terms of practical execution and degree of structuring. Chapter 7 provides explanations along with a comparison of three process improvement methods: Kaizen, DMAIC, and the Five Focusing Steps. The practical managerial implications and brief guidance for choosing and co-using them are addressed.

This whole book, including the Constraint-Driven Lean Six Sigma concept, suggestively implies that elevating one critical point of the operations, namely, the bottleneck, is not enough. To maximize flow and fully exploit the bottleneck, a stable and predictable flow in all other processes is indispensable. The operation system substantially needs reduced variability in all the processes, which is strictly related to systematic waste elimination. The flow of operations has to be predictable, responsive, and work according to the rhythm.

Chapter 8 addresses practical solutions supporting steady and synchronous flow of operations. The flow management techniques, namely, Kanban and Drum-Buffer-Rope, are discussed. Both are characterized, and a fundamental comparison is provided. Suggestions in terms of the application are derived, including fields of advantages.

This book tries to address the important gap in the management literature regarding the joint use of Lean Six Sigma and the Theory of Constraints in an organization's pursuit of excellent operations. The appropriate combination of elements from these methodologies can create a smart and robust system of process changes, which eventually leads the organization toward a favorable development trajectory. This book places equal emphasis on delivering inspiring concepts and comprehensive explanations of key issues, as well as empirical evidence and business practical cases. The authors hope that it will be an inspiring and useful resource.

2

Fundamentals of Lean Management, Six Sigma, and Theory of Constraints

2.1 General Concept

Companies usually have a great experience of success and failures in different improvement initiatives, cost-saving programs, quality programs, new organizational systems implementations, and so on. On the other hand, a variety of approaches, methods, and methodologies dedicated to making organizational processes work better might place managers in front of real dilemmas and quandaries about what to apply to achieve the most efficient outputs, which means a high-performance and highly competitive organizational system. The essential role of the three highly effective management methodologies in process improvement is presented in terms of their origins, basis, fundamental assumptions, and principles. Along with these, the main issue is how to apply methodologies in a company, which means what is essential for their proper operation within the organization.

Lean Six Sigma has emerged as the integration of two previously separate approaches. To understand it, it is indispensable to know how they started. Lean Management originated in the post-WWII Japanese economy. Japanese companies began improving their operations and built a very original management system covering the entire organization and cooperation networks. A few decades later, American companies developed a system for improving organizational processes based on statistical metrics and also aiming at process excellence; this is Six Sigma. A very interesting concept covering all company operations was also developed by the physicist Goldratt under the name of the Theory of Constraints. This theory offers a simple rule on how to improve company profit.

Each of the three concepts has its own specific management principles. In some, they are clear and legible; in others, more covered up. However, all three target the same goal: increasingly efficient organizational processes that contribute to the success of the entire company. Additionally, the three approaches share very similar crucial organizational requirements to function efficiently within the organization, and one of these is of primary importance.

 DOI: 10.1201/9781003655237-2

Among many convergent and similar elements, one issue is of special importance. As proven by empirical study, employee engagement is indispensable for achieving significant improvement of processes measured by typical specific indicators. Engaged people lead to the effective carrying out of improvement projects, and the developed solutions are permanently adapted into the daily routines of organizational processes.

2.2 Introduction

There are many ways to improve working processes in organizations: making them less costly, more flexible and scalable, consuming fewer resources of any kind—especially workers' effort—and ultimately delivering more value to customers. Which manager would not want these benefits? These needs have motivated managers for a very long time, since the first ideas and methods of process improvement were introduced in companies, in search of effective methods and methodologies in this field. Since the 1940s, many interesting concepts have been developed, some of which are of special interest because of their usefulness and efficiency in today's business organizations.

Lean Management and Six Sigma are among those that offer effective changes in the functioning of organizational processes, provided, however, that they are used persistently and correctly. The Theory of Constraints is, in fact, a global operations management theory based on one fairly simple rule. It offers management principles and methods that might lead to successive changes in the functioning of processes within the organization. Each of these approaches must be critically reviewed, considering their critical points, as they may have great potential for mutual complementarity and synergy. They must first be analyzed at the level of their origins, general management principles, and the most important success conditions when using them.

This chapter aims to incrementally present the fundamentals of three process improvement methodologies: Lean Management, Six Sigma, and Theory of Constraints. This chapter also presents an empirical case study based on a survey of 380 medium and large companies, which demonstrates the most important condition for the successful use of these methodologies.

2.3 Origins Giving the Understanding

Lean Management is one of the most valued management methods, with numerous business successes confirmed by numerous companies and researchers. Lean Management is a complex and multidimensional methodology. The literature argues that since its inception Lean has been developed as an interdisciplinary approach encompassing operations management, organizational behavior, and strategic management. According to Krafcik (1988), a pioneer of this issue in Western management, Lean in briefly defined as the use of fewer organizational resources compared to traditional mass manufacturing. This author emphasizes the high productivity effect characteristic for Lean, which is arguably the most important factor in its attractiveness for managers interested in this approach.

However, Lean Management has been extensively researched and explained by many recognized authors worldwide. Womack et al. (1990), the authors who built an important framework for today's understanding of Lean and laid the foundation for its spread, understand this management approach as a combination of mass production and craftsmanship, based on accumulating the best principles and practices aimed at continuous improvement. Other popular brief explanations in the field describe Lean as the reduction of waste, a socio-technical system of delivering value to customers, and a management philosophy.

Emiliani's (1998) explanation of Lean Management boils down to the everyday attitudes and behaviors of employees within an organization. This author developed the concept of "Lean behaviors," which he understands as behaviors that add and create value within an organization. Lean behaviors lead to the minimization of waste associated with typically occurring activities such as contradictory actions, poor cooperation, and negative attitudes. Lean behavior is the opposite of irrational and confusing information created by employees at various levels, which results in delays or interruptions in work. This explanation of Lean Management raises the key issue that the efficiency of the daily behavior of all employees has a direct impact on overall production efficiency and the satisfaction of customers and other stakeholders.

However, a comprehensive understanding of Lean Management absolutely requires consideration of the Toyota Production System (TPS), which contributed the basis of Lean Management, further principles and tools, and to this day is still the unsurpassed Lean standard. Liker (2004) underlines that in the Toyota system, the organizational tools and management principles used work together to support continuous process improvement by employees. The success of TPS stems from the engagement of employees in continuous improvement, leading to waste reduction, which in turn leads to a sustainable competitive advantage.

As a great reference object, Toyota factories, with their methods, principles, and results, have been the object of consultants' and researchers' interests. One spectacular observation was made by Cusumano (1988), who compared the productivity results achieved by Toyota's system to its main competitors. The comparison time span started from the early beginning when Toyota built its system, since the 1950s. The author calculated the adjusted productivity index as manufactured cars per employee; the outputs were corrected for subcontracts and also standardized. In 1965, with an average index for American car manufacturers of 1, the index for Nissan plants was 0.9 and for Toyota 1.5; while in 1985, again with a productivity value of 1 for American manufacturers, the ratio for Nissan was already 1.9 and for Toyota as high as 2.2 (Cusumano, 1988, p. 30). The ratio of 1 to 2.2 shows the huge gap in productivity achieved by Toyota plants. The data show how Japanese car manufacturers managed to double their productivity in just 20 years, still having virtually the same product designs and identical technologies as Americans at the beginning and the end of the time span.

Lean Management can be summarized as a management system that aims to optimize processes by eliminating waste through process improvement in order to deliver value to the organization's customers (Sinha & Matharu, 2019). There is no doubt that the "Toyota method" has a special place in Lean management; Toyota's system is considered a benchmarking model and still inspires thousands of companies. In fact, Toyota was to some extent inspired by Henry Ford's flow production system (Ohno, 1988). It is clear that Lean methodology is largely derived from the Toyota system, whereas the foundation of Six Sigma was developed by Japanese electronics manufacturers.

Six Sigma has its origins in the 1970s, when the Japanese took over the production of Quasar televisions in the United States from Motorola. It turned out that the Japanese, using the same workforce, the same technology, and the same designs, radically reduced the number of manufacturing defects in a very short time, so that the number of defects was 20 times lower than under Motorola's management (Pyzdek & Keller, 2010, p. 4). At that time, American electronics were losing out to Japanese products. Americans became convinced that the cause of problems with the quality and competitiveness of products in their plants lay in the area of management.

Six Sigma, initially referred to as the 6 Steps of Six Sigma, was developed at Motorola's plants in response to the loss of market share to Japanese competitors. The leitmotif of the Six Sigma methodology is the statistical description of the variability that occurs in the processes taking place in an organization. A special place is occupied by the statistical parameter of standard deviation, which in statistics determines the dispersion of a data set relative to the mean value of that set. However, the approach also integrated elements of Toyota's Lean Manufacturing, along with inspirations from authors such as Joseph Juran, whose works laid the foundations for Total Quality Management (TQM). Six Sigma gained popularity when large American companies, such as General Motors, began to use it. The approach evolved to focus not only on the traditional understanding of quality, that is, compliance with defined requirements, but also on increasing profits by improving efficiency and value for customers (Pyzdek & Keller, 2010).

The term Six Sigma, which refers to statistical six sigma values, that is, six standard deviations, implies that the process should operate within six standard deviations. This means that the range of acceptable variability is very narrow, which in turn means that only minimal defects and errors occur in processes. Six Sigma aims for near perfection, meaning that the process should generate no more than 3.4 defects per million opportunities, resulting in a process performance rate of 99.99966%. This level of perfection is achieved by controlling and managing process variability using statistical methods. The significance of this challenge can be illustrated by the benchmark, which is the state of typical processes in organizations not practicing process improvement. According to sources, it can be approximated as a three-sigma level, which was also adopted by the traditional quality management paradigm, meaning that there are as many as 66,800 defects per million opportunities in the process (Mehrjerdi, 2011, p. 79). The difference in the number of defects is really huge.

The Six Sigma methodology is based on a very clear statistical concept: the aim is to reduce process variability, thereby minimizing errors and defects in organizational processes. The overarching goal of this methodology is to increase customer satisfaction and reduce costs associated with low quality (Six Sigma, 2018). Six Sigma is based on the statistical descriptions of processes within an organization; that is why statistical techniques are fundamental to this approach. Statistical tools are used to analyze process data, including the determination of statistical relationships between variables of many types. Among the popular techniques in Six Sigma are measures of dispersion and location related to various variable distributions, analysis of variance, process characteristic control charts, statistical tests such as chi-square, correlation coefficients, regression analysis, and basic process capability indicators (Voehl et al., 2013).

Based on these two mentioned methodologies, the Lean Six Sigma approach emerged. Lean Management does not have enough methods for introducing statistical control into processes, while Six Sigma does not consider improvements in process speed or significant reductions in capital invested in the business (George, 2002, p. V). Therefore, in order to achieve fast and noticeable results, it is necessary to integrate both approaches. Lean Six Sigma is a management methodology that maximizes value for stakeholders by quickly improving customer satisfaction, reducing costs, improving quality, improving process speed, and reducing invested capital. It works on the basis of two main assumptions: reducing variability, typical of Six Sigma, and eliminating non-value-adding activities, which is characteristic of Lean Management (Patel & Patel, 2021, p. 633). Both approaches complement each other in Lean Six Sigma, making it an effective tool for improving quality and productivity in various sectors, including manufacturing, as well as in various types of services such as healthcare and higher education.

The Lean Six Sigma methodology is gaining popularity among companies. It appears that the synergy of both approaches allows for the effective solving of various problem types in organizations, making Lean Six Sigma a comprehensive and tool-rich approach to everyday process improvement. An important advantage is its data-driven approach to problems and process issues; Six Sigma provides appropriate tools and knowledge to apply advanced data analysis. Accurate identification and resolution of problems are assured. The focus on customer satisfaction is at the heart of Lean Six Sigma; it is perceived as the most important result of this system. Satisfaction is achieved by improving the quality of products and services while reducing order fulfillment times (Antony et al., 2017). Lean Six Sigma leads to committed leadership and employee engagement in process changes. This methodology offers effective change management and real, positive changes in organizational processes.

A slightly different perspective is offered by Goldratt's Theory of Constraints, developed in the 1980s. Initially, the theory functioned as a system aimed at improving production efficiency, but over time its application was extended to all areas of organizational management. A key element of the theory is the assumption that every organization must have some constraint that currently acts as a barrier to its development, and that identifying and properly managing this constraint leads to improved productivity and profitability (Rahman, 1998).

According to the author of this concept, Goldratt, a constraint is that limits the system from achieving higher performance in relation to its goal (Goldratt, 1988, p. 453). A fundamental aspect of the Theory of Constraints is the belief that a constraint is not negative in its nature; on the contrary, a constraint should be perceived as an opportunity for improvement and development of the organizational system. Thanks to this perspective, the Theory of Constraints provides the idea of continuous development of an organization by eliminating, one by one, the constraints that stand in the way of the goal for which the organization exists.

According to the Theory of Constraints concept, any organization is a complex system in which various elements, organizational processes, and all their components interact with each other, performing their work sequences with certain throughputs. One element of the entire system, characterized by the lowest throughput, determines the rhythm of the entire system's functioning; this is a constraint, or, in other words, a bottleneck of the system. A constraint in an organization can be any factor that limits the organization's ability to achieve better results in relation to its goal, that is, to achieve successive stages of development and growth. Limitations can be of various kinds and relate to many different areas; they can be physical or managerial in nature, meaning they can result from physical resources or from non-physical factors, including the policies pursued by managers (Rahman, 1998).

The Theory of Constraints opposes the idea that managerial decisions are subordinated to accounting principles, understood in a traditional way. Therefore, it proposes three key measures that describe the efficiency of an organization's operating system: throughput, investment, and operating costs. Throughput is referred to as the rate at which an organizational system generates money from the sale of manufactured goods and delivered services. Investments include equipment and buildings, but also inventories, which are expenditures on items that the organization intends to resell.

Operating costs are the funds that the organization must spend to transform purchased inventories into products that will be sold. These three measures create a simplified financial model of the organization. The theory assumes that any activity in an organization that does not positively affect at least one of these three measures is unproductive, that is, disadvantageous (Bhardwaj et al., 2010).

The Theory of Constraints concept, along with the framework methodology behind it, is closely related to Lean Management, as both aim to optimize production and operational efficiency (Rahman, 1998). Main authors in the field emphasize that the ultimate goal of the Theory of Constraints is to support a culture of continuous improvement by continuously identifying and resolving constraints, which leads to better results achieved by organizations (Cox & Schleier, 2010, p. XXXII). Scholars even state that it describes the entire operational management, primarily because the Theory of Constraints avoids the pitfalls of local optimization, offering instead an integrated approach that encompasses all functions, processes, and resources of the organization. It also serves to increase the efficiency of the entire system (Gupta & Boyd, 2008). These characteristics give the Theory of Constraints a unique potential to deliver significant benefits for process improvement.

2.4 Principles—Basis for Applying Methodologies

For effectively running Lean Six Sigma as well as the Theory of Constraints, more precise and concrete issues are necessary. Such management principles form the foundation and framework in the organization, opening a space to practically operate the methodologies. Lean Management is one such methodology, which abounds with substantial management principles derived from the Toyota system by investigators observing this system. These principles are meaningful and support the organization on the route toward process improvement.

The five Lean Management principles by Womack and Jones (1996) are probably the most popular and most frequently cited explanation of what is practically the most important for Lean application by a company. The authors presented them as basic principles that enable companies to eliminate waste, improve customer value, and achieve higher revenues along with improved competitiveness.

The first Lean principle (1), derived by Womack and Jones (1996), is about understanding the customer by the organization, particularly what exactly brings value in the company's products to customers. In fact, customer perceptions and expectations are bent over backwards by marketing theories. However, here this principle is meant to be very comprehensive, in-depth, practical, and serving other system elements. Definitely, this principle is not about generalities or wishful thinking; every product delivered by an organization should have deeply understood value in terms of how it is perceived by customers. Defining value can be difficult, as the concept of value can be interpreted differently within the organization itself, and it is heterogeneous in terms of customer structure. Therefore, this principle requires that sufficient attention and time be devoted to studying, observing, and understanding the value exhaustively.

The organization must identify the internal pathways through which the value-creating stream passes for each of the products the company delivers (principle 2). The value stream should be understood as all activities in the company necessary for the creation of a product, which requires going through three key stages: product preparation—from idea to production launch; order acceptance and shipment—from order acceptance to delivery; and physical transformation—from raw materials to the finished product at the customer's site (Womack & Jones, 1996, p. 141). Identifying the value stream allows the organization to recognize waste.

The accurate identification and understanding of the value stream allow for the organization of a smooth flow of value-creating activities (principle 3). This principle means organizing a continuous workflow for each product or project, without downtime, waiting, or other waste. A fluent value stream covers all the activities from end to end, that is, from the entry of materials to the shipment of the finished product. Achieving smooth and continuous flow often requires significant reorganization of machinery, coordination methods, plant layouts, and many other operational issues.

Authors of the five Lean principles recognized the great importance of determining demand before producing and delivering goods or parts, performed in short cycles, as in the widely known and admired Toyota Just-in-Time system. This is a pull-based coordination system (principle 4). This principle states that you should produce and deliver exactly what the customer wants and when they want it. The flow should be organized in such a way as to allow customers to "pull" products from the organization's value stream, which eliminates many types of waste, including excess inventory and excessive volumes of production.

The last of the five principles can be expressed as for excellence (principle 5). This is possibly the most powerful principle and, at the same time, the most difficult to implement, as it primarily depends on employee behaviors and attitudes. In fact, it is not possible to implement the above four principles as a one-time action, so it is necessary to pursue continuous process improvement, starting with defining value, flow, eliminating waste in the flow, and improving the pull system. Permanent process improvement leads to a reduction in effort, time, space, and costs, as well as a decrease in the number of errors, while delivering products that increasingly meet the actual needs of the customer (Womack & Jones, 1996, p. 141).

These five Lean principles were observed and described as the first explanations of what Lean Management is all about. They clearly demonstrate the fundamental originality of Lean within the context of a typical mass production organizational system, as traditionally understood in the 1990s. They are still fully relevant and provide support for those who want to follow the Lean development road. Liker (2004) provided a more detailed list of principles of Toyota, called the Toyota Way. The list consists of 14 principles, organized into four groups and arranged in the shape of a four-level pyramid.

At the base of the pyramid there is only one principle, which forms the foundation of the entire system elaborated by Liker; this principle is referred to as "philosophy." This principle states that all management decisions should be based on a long-term philosophy of actions, even if it proves necessary to sacrifice short-term profits for this purpose (Liker, 2004, p. 86). The principle describes the need to prioritize long-term thinking in business decision-making. It is crucial to develop a philosophical sense of purpose that transcends short-term decisions. The overarching goal is to create value for the customer, society, and the economy. This principle entails demonstrating patience, focusing on long-term results, investing in people, products, and facilities, and maintaining an uncompromising commitment to quality.

The second level of the Toyota Way pyramid consists of the next seven principles, grouped under the heading "process." The group refers to the need to focus attention and devote efforts to the proper functioning of processes in the organization. It is necessary to create continuous flows in processes so that any problem present in them become visible at once. In the organization of the flow, pull logic should be used to avoid overproduction in processes. The next tips are for balancing work in processes, that is, arranging production tasks in repetitive sequences, as well as stopping the entire process immediately when a problem is detected. Standardized work, that is, work based on established sequences of activities, is necessary. Process control and coordination of work have to base on visual techniques, easy-to-understand by performers.

The next higher group of principles on the pyramid concerns the development of the organization's employees and business partners. First of all, there is a need to develop leaders within the organization who understand the organization's philosophy well and teach others. It is also necessary to build teams consisting of exceptional employees who are constantly developed in accordance with the organization's philosophy. The last, third principle in this group refers to building a network of partners for the organization based on the principles of trust, including suppliers, who should be helped to improve their organizational system.

At the very top of the Toyota Way pyramid is continuous problem-solving. Primarily, problems in the flow of processes and ways of solving them should be judged where they occur, that is, where the process takes place. This means that the organization's top management should systematically visit the places where the basic processes take place. Decisions in the organization should be made slowly, with deep consideration and by consensus. Such a decision-making approach is potentially of great importance, in contrast to impulsive and interest group-based decisions, which are often influenced by internal and external image factors. And finally, at the top of the top Toyota Way pyramid is continuous organizational learning (still the fourth layer of pyramid). This is about knowledge sharing, training and motivating people by line managers, getting experience from mistakes and defects, and above all form improvement in processes performed by teams.

The Toyota Way is totally in line with abovementioned Womack's and Jones' five principles. A set of management principles taken directly from Toyota's plants is an important guide to Lean Management and precious source of inspiration for managers and consultants. Liker's studies are also abundant with a number of management methods developed by Toyota, describing practices that can be directly imitated by others or developed according to the needs of individual organizations.

Among the crucial elements of Lean Management system, one more that should be noted as fundamental is the kind of widely practiced routine for performing improvements by employees in processes they are working in. This everyday practice of continuous improvement is referred to as Kaizen by many practitioners. Kaizen is an important Japanese idea that has a direct reference to the continuous improvement of processes in an organization; it delivers fundamental inspiration for continuous improvement. Hence, the Kaizen method or system is often referred to as a philosophy. Authors point out that Kaizen is at the heart of all management methodologies focused on process improvement, such as Lean Management, TQM, and others (Suarez-Barraza et al., 2011).

Kaizen might even be treated and practiced as an entire methodology for process improvement, equipped with numerous supporting tools, focused on processes and people, continuous in nature, and based on broad employee participation. Then this could be in itself a formula for business excellence in an organization, but focused primarily on excellence in managing people and other organizational resources (Macpherson et al., 2015). However, the wide Lean Management methodology proposes several methods for Kaizen organizational routines, as A3 Report and many versions of Kaizen workshops and others. The organization needs some method or methods widely practiced within to perform daily process improvement—a Kaizen routine. If not, the Lean or other methodology in fact does not exist.

Six Sigma is not so clearly specified in terms of general management principles; this is a different management methodology, much more focused on operations but less on the whole company management issues. It also has to be noted that Six Sigma was developed as a management system pretty transparent for commentators and researchers, not like Toyota's system which was growing for decades not studied by external observers. So, the general perception of Six Sigma has been much more fitted into the entire Western system organization management. However, also Six Sigma is characterized by some general and broad principles.

First of all, the leitmotif of whole Six Sigma approach is mastering all processes variabilities, occurring defects, mistakes, discrepancies, and reducing any unpredictability and uncertainty. This is mathematically defined perfection of processes, all kind of processes, the manufacturing and service ones, main processes and supporting. Such an approach in organization is based on data and facts so that accurate and reliable measurements of process characteristics are obvious in Six Sigma. The universal goal in this system is to reduce the variance of variables characterizing processes in the organization. And aiming at perfection means to struggle to reach the level of six standard deviation and beyond, if justified and if possible.

The consequence of the principle to mastering process variabilities is to provide statistical control to processes (e.g. Statistical Process Control). This methodology has a long and rich history; it offers many useful techniques based on the rules of statistical inference. Six Sigma is characterized by a strictly statistical and analytical approach to decision-making. It uses advanced statistical tools to understand process variability and identify the root causes of problems.

The second fundamental principle of Six Sigma can be considered the voice of the customer. The essential aim of the Six Sigma system is to provide goods and services that meet or exceed customer expectations. Many techniques are devoted to systematic gathering of data, studying, quantifying, valuating, and concluding in terms of expectations and perceptions of customers. Six Sigma combines the knowledge obtained from customers with measures and statistics in conjunction with process improvement methods to increase customer satisfaction, ultimately increasing profit, customer retention, and their advocating behaviors (Six Sigma, 2018, p. 15).

The notion of quality in fact bases ultimately on what is desired by the customers. The system of Six Sigma assumes to perform investigations of many kinds to define the voice of the customer, questionnaire surveys, focus groups, feedback forms, social media examination, and other. Apart from proactive voice of the customer companies, many frequently used in Six Sigma methods have entered the search for the customer's voice. In the Quality Function Deployment (QFD) method, popular in the system, the voice of the customer is identified by establishing a list of product features that are important to customers, along with an assessment of the significance of each feature using a scale. The Kano model concept is employed to support recognition of the voice of the customer and deeper understanding of the complexity and dynamics of the customer experience with products.

However, the fundamental of Six Sigma is formed by two specific routines, one of performing improvements in processes and another of designing process and products, both very similar to each other and both very rich in methods—most of all in statistical methods. Both DMAIC and DMADV consist of five steps. The standard procedure for process improvement, similar to process/product design, starts with the definition of the real problem that should be solved or the issue should be designed. This is an important issue for the final success; without clear determination of aims, scopes, and results frame, the output will surely be questionable as to its value. The first step of "definition" delivers schedule of future works and risk factor evaluation.

The next two steps of the improvement routines are on measuring and analyzing. These are about gathering the values of many indicators; it is necessary to give meaning and understanding to the indicator values; the key point of the routine is to search for the real causes shaping the problems being solved. When designing new processes or products, plenty of techniques are also employed; most of them are devoted to quantification of requirements and/or product characteristics. The value in the eyes of the customer is repeatedly specified and confirmed.

The last two steps in the routine dedicated to process improvement are "improve" and "control." The fourth step consists of implementation of the solution discovered in the previous steps into the process real work, which should fully answer the problem defined on the first stage. The final (fifth) step is on monitoring the effects achieved and to set standards assuring sustainability of changes and spreading them to other processes.

In the routine devoted to process and product design, the two last steps are "design" and "verify." The "design" step involves developing a high-level design; it means defining detailed solutions and specifications, all according to hitherto analysis. Predictive models, simulations, prototypes, or trial runs are employed to verify the accuracy of detailed design solutions in the context of achieving previously set goals. The final (fifth) verification step is on assessing whether the designed product or a process is acceptable to all stakeholders, conducting pilot tests, confirming expectations, and documenting conclusions. Final versions of Standard Operating Procedures are developed.

The basic improvement formula is an organizational code permeating the entire Six Sigma system and building it. This basic method is fully based on project management rules; it is formalized and bureaucratic in nature. Six Sigma principles, along with a mastering variability, the focus on value to the customer, and elementary improvements according to DMAIC show to some extent different face of organizational improvement however not contradictory to Lean Management approach but rather consistent and complementary.

When considering the Theory of Constraints, there are no management principles explicitly elaborated by the literature; however, the main issue for this approach is to be focused on the goal, as the main novel explaining the theory by Goldratt (1984) is entitled. The organization's main goal is to increase the throughput and thanks to this assure high profit now, and in the future, as according to Goldratt the organization is existing for this. While maximizing the throughput the company should simultaneously reduce both inventory and operating expenses. The management principle to increase throughput is the one that permeates and defines the entire approach of Theory of Constraints. The throughput accounting developed within this approach is subordinated to this purpose.

Another matter that rises to the rank of a cross-cutting principle for the organization following the Theory of Constraints is looking for the bottleneck of the system and then increasing the capacity of the bottleneck point of organizational system. Systematic dealing with a system's bottlenecks one by one is recognized as a way to fulfill the goal of increasing throughput. The issue of how to recognize an organization's constraints (bottlenecks) along with the methods of dealing with them is widely discussed by scholars and professionals. This belief, that each system has at least one constraint, if not the system it would have unlimited throughput, so the improvements have to be focused on the constraint is even a model approach to Systems Thinking, as suggested by many scholars.

There is one more general organizational principle to follow when aiming at the practical use of Theory of Constraints, which is to organize a smooth flow of processes clocked by the rhythm of bottleneck bandwidth. There is a designed dedicated method to organize such a flow—Drum-Buffer-Rope. When the entire flow is based on the bottleneck rhythm it effects with maximum throughput and leads to increased profits. The principle of organization of the entire flow according to the bottleneck capacity is in line with Lean Management principles referred to value stream explained above.

2.5 The Central Role of Engagement—Empirical Case Study

A common denominator of all improvement methodologies is a culture of continuous improvement. This frequently used phrase in the field means the idea that employees are empowered to initiate and implement changes to eliminate waste (Alnadi & McLaughlin, 2021). Such a culture fosters an organizational learning environment, as only then are employees effectively encouraged to contribute new ideas. This environment assures them that they will not be blamed for any mistakes they make (Loh et al., 2019). As a result, such a culture contributes to improved quality and efficiency within the organization (Alnadi & McLaughlin, 2021).

The essence of the organizational culture of continuous improvement is the visible engagement of a wide range of employees. From a managerial practical viewpoint, employee engagement might be seen as the other side of the same coin. Employee engagement in process improvement means that the commitment and intelligence of ordinary employees are devoted to internally inspired initiatives, with a balance between the quantity and quality of work, which is beneficial to employees (Singh & Singh, 2015, p. 76).

Process improvement is mainly based on evolutionary incremental changes. These kinds of changes can transform the processes as a whole only when all employees, especially front-line employees who have executive roles in processes, contribute to them. The authors emphasize that employee engagement is essential for continuous improvement, primarily because employees who are closest to the problem, that is, the workflow and process flow, are also in the best position to make decisions aimed at improving them (Wickramasinghe & Chathurani, 2021, p. 888).

Employee engagement is studied in many contexts, as it is a vivid management concept in general. The Gallup Institute has particular expertise in investigating employee engagement in various types of organizations. According to Gallup, achieving employee engagement is linked to the ability to win the hearts and minds of employees in order to instill in them an inner desire and passion for excellence (Fleming & Asplund, 2007, p. 2). Although the spiritual element seems irrational and difficult to investigate, this approach to engagement is consistent with the latest research, which emphasizes that understanding employee engagement should consider employees' mental states, and in particular, the presence of passion for work as a key element.

In fact, what is needed for process improvement in the organization is employee engagement as proactive behavior supported by their passion and the use of their full creative thinking potential and experience. The issue of employee engagement, along with the adaptation of improvement methodologies in companies, has been investigated in a survey of medium and large companies located in Poland. This country's business ecosystem is quite internationalized and competitive; companies, especially larger ones, already have great experience using Lean, Six Sigma, and other similar concepts.

The sample of investigated companies consisted of 380 entities, which met the requirements for a statistically representative sample size for the country for medium and large companies employing 50 or more employees. The survey was performed using the Computer-Assisted Web Interviewing (CAWI) methodology. Respondents in the companies were top and middle managers who declared their experience and knowledge in process improvement affairs in organizations.

In the survey, information regarding employee engagement levels and improvement methodologies in companies was gathered. Additionally, typical operating results characteristic of process improvement were approximated by surveyed managers. It is assumed that having noticeable operating results in the organization is a sign that process improvement truly exists and functions properly and effectively. The research tool included several questions designed to capture these issues, scaled with a Likert scale; the use of process improvement systems was scaled as yes or no.

In the presented short version of employee engagement six queries, the first three (Eng 1 to Eng 3) try to capture typical employee behaviors characteristic to process improvement, while the next three (Eng 4 to Eng 6) determine more general, typical to engagement of employees, issues. A few of the queries were inspired by Gallup's GWA tool (Harter et al., 2002) used for years for investigation of workforce engagement. The list of these queries is presented below along with the codes assigned to them.

1. Our employees are personally engaged to their work and its improvement, Eng 1.
2. The quality of production and/or services is important to our employees, Eng 2.
3. Employees often suggest ideas for improvements in their workplaces, Eng 3.
4. Our employees like their jobs, Eng 4.
5. Employees speak well of their work and the company, Eng 5.
6. Employees identify themselves with goals and values of company, Eng 6.

The understanding of employee engagement importance is possible when confronting it with typical results that organizations achieve when practicing process improvement. The literature elaborates several result indicators achieved by companies when conducting real improvements in business process (Rother & Shook, 2009; Ohno, 1988). One of the primary outcome measures is productivity increase; many operational measures referred to time in processes are bettering, quality is increasing, and many of the waste diminishing. In the investigation questionnaire, respondents were asked whether their companies noted positive changes in the last three years in the following indicators:

1. The productivity of our processes has improved, Res 1.
2. Process lead times (LT) and the entire value stream have been shortened, Res 2.
3. We have reduced cycle times at workstations (C/T) and changeover times (C/O), Res 3.
4. The level of inventory in the company has decreased, Res 4.
5. The level of quality in processes has increased, Res 5.
6. Our customer satisfaction has improved, Res 6.
7. We have eliminated many types of waste and achieved savings, Res 7.
8. We have improved flexibility in our processes, Res 8.
9. We have reduced our unit operating costs, Res 9.

The Spearman's rank correlation coefficient R has been calculated to determine the relationship between variables representing employee engagement and the operational results achieved by companies. The output is shown in Table 2.1.

The R coefficients are statistically significant for all pairs of variables in Table 2.1. All p-values are zero to three decimal places, so they are not

TABLE 2.1

The correlation between employee engagement and operational results

	Eng 1	Eng 2	Eng 3	Eng 4	Eng 5	Eng 6
			***R* coefficient, $p < 0.05$**			
Res 1	0.470	0.400	0.397	0.464	0.475	0.417
Res 2	0.442	0.475	0.391	0.477	0.482	0.465
Res 3	0.346	0.350	0.356	0.429	0.432	0.429
Res 4	0.209	0.207	0.157	0.323	0.305	0.297
Res 5	0.503	0.515	0.443	0.503	0.500	0.478
Res 6	0.477	0.471	0.421	0.508	0.511	0.525
Res 7	0.415	0.395	0.373	0.479	0.442	0.447
Res 8	0.465	0.436	0.365	0.486	0.507	0.503
Res 9	0.360	0.390	0.336	0.475	0.492	0.462

presented. The *p*-values, along with the *R* values, firmly imply that there is a statistically significant relationship between all variables of employee engagement (Eng) and all variables approximating the operating results (Res) declared by the studied companies. The values of *R* are relatively high compared to values frequently observed in studies in this research area. The maximum *R* score is 0.525, while the median is 0.443; according to statisticians, an *R* value above 0.4 should be interpreted as a moderate level of relationship.

Hence, there is strong proof that when employee engagement is at a higher level in a company, such a company also achieves better operating results. If a company achieves good results in operational indicators, this means that process improvement is working in an appropriate manner. However, there is also another interesting issue: whether an organization that uses a process improvement methodology also achieves higher results than others. Spearman's correlation coefficient has been calculated for pairs of variables describing operating results and the declaration of using the three improvement methodologies. For such scaled variables, the *R* coefficient is appropriate. The statistical data is shown in Table 2.2. LM means Lean Management, SS means Six Sigma, and TC means Theory of Constraints.

Not all pairs of variables appeared statistically important; the *p*-value was greater than 0.5, so these coefficients are not presented in Table 2.2. From a statistical point of view, each methodology does not have a relationship with all types of results considered in this research. In each case of improvement methodologies, there are some result variables where *R* coefficients are not statistically significant. The range of statistically important *R* values is between 0.101 and 0.214, with a median of 0.139. In statistics, such levels of *R* are interpreted as very weak relationships between variables. Thus, according to the research data and statistical calculations, the improvement methodologies have a very weak or even no relationship with the operational results obtained by organizations.

TABLE 2.2

The correlation between improvement methodologies and operational results

	LM		SS		TC	
	R coef.	*p*	*R* coef.	*p*	*R* coef.	*p*
Res 1		-		-		-
Res 2	0.108	0.0355		-		-
Res 3		-	0.154	0.0026	0.113	0.0270
Res 4		-	0.119	0.0208	0.137	0.0076
Res 5	0.170	0.0009		-	0.107	0.0374
Res 6	0.157	0.0022	0.101	0.0486	0.140	0.0062
Res 7	0.214	0.0000	0.115	0.0253	0.136	0.0077
Res 8	0.168	0.0010	0.102	0.0469	0.172	0.0008
Res 9	0.188	0.0002		-	0.165	0.0012

Now, it is necessary to look at two sets of calculations presented above. In the case of employee engagement variables, all of them are pretty firmly correlated with operational results (Table 2.1). At the same time, the improvement methodologies variables are very weak or not even correlated with operational results declared by the investigated companies (Table 2.2). From this emerges an absolutely important conclusion: as improvement methodologies offer a very good way for incessant positive changes in organizational processes, they absolutely require organizational struggles and efforts toward real engagement of employees in the organization. If not, the process improvement is treated superficially by the organization and eventually does not bring large and stable operational results.

While following improvement methodologies, it is fundamental for responsible managers in companies to make profound changes in the social sphere of their organizations. They should assess whether there are internal obstacles to company-wide employee engagement in process improvement. While advocating improvement philosophies and implementing new methods and tools, they should primarily focus on eliminating barriers to individual engagement, as this is usually a longer process and such obstacles often exist in companies. Neglecting this matter will surely lead to the failure of continuous improvement in a company.

References

Alnadi, M., & McLaughlin, P. (2021). Critical success factors of Lean Six Sigma from leaders' perspective. *International Journal of Lean Six Sigma*, 12(5), 1073–1088.

Antony, J., Snee, R., & Hoerl, R. (2017). Lean Six Sigma: Yesterday, today and tomorrow. *International Journal of Quality & Reliability Management*, 34(7), 1073–1093.

Bhardwaj, A., Gupta, A., & Kanda, A. (2010). *Drum-Buffer-Rope: The technique to plan and control the production using Theory of Constraints*. World Academy of Science, Engineering and Technology, 69.

Cox, J. F., & Schleier, J. (2010). *Theory of Constraints Handbook*. McGraw-Hill Education.

Cusumano, M. A. (1988). Manufacturing innovation: Lessons from the Japanese auto industry. *Sloan Management Review*, 30, 29–39.

Emiliani, M. L. (1998). Lean behaviors. *Management Decision*, 36(9), 615–631.

Fleming, J. H., & Asplund, J. (2007). Where employee engagement happens. *The Gallup Management Journal*, November.

George, M. L. (2002). *Lean Six Sigma*. McGraw-Hill.

Goldratt, E. M. (1984). *The Goal: A Process of Ongoing Improvement*. North River Press.

Goldratt, E. M. (1988). Computerized shop floor scheduling. *International Journal of Production Research*, 26(3), 443–455.

Gupta, M. C., & Boyd, L. H. (2008). Theory of constraints: A theory for operations management. *International Journal of Operations & Production Management*, 28(10), 991–1012.

Harter, J. K., Schmidt, F. L., & Hayes, T. L. (2002). Business-unit level relationship between employee satisfaction, employee engagement, and business outcomes: A meta-analysis. *Journal of Applied Psychology*, 87(2), 268–279.

Krafcik, J. F. (1988). Triumph of the lean production system. *MIT Sloan Management Review*, 30(1), 41.

Liker, J. K. (2004). *The Toyota Way: 14 Management Principles from the World's Greatest Manufacturer*. McGraw-Hill.

Loh, K. L., Mohd Yusof, S., & Lau, D. H. C. (2019). Blue ocean leadership in lean sustainability. *International Journal of Lean Six Sigma*, 10(1), 275–294.

Macpherson, W., Lockhart, J., Kavan, H., & Iaquinto, A. (2015). Kaizen: A Japanese philosophy and system for business excellence. *Journal of Business Strategy*, 36(5), 3–9.

Mehrjerdi, Y. Z. (2011). Six-sigma: Methodology, tools and its future. *Assembly Automation*, 31(1), 79–88.

Ohno, T. (1988). *Toyota Production System: Beyond Large-Scale Production*. Productivity Press.

Patel, A. S., & Patel, K. M. (2021). Critical review of literature on Lean Six Sigma methodology. *International Journal of Lean Six Sigma*, 12(3), 627–674.

Pyzdek, T., & Keller, P. A. (2010). *The Six Sigma Handbook: A Complete Guide for Green Belts, Black Belts, and Managers at All Levels* (3rd ed.). McGraw-Hill.

Rahman, S.-u. (1998). Theory of constraints: A review of the philosophy and its applications. *International Journal of Operations & Production Management*, 18(4), 336–355.

Rother, M., & Shook, J., (2009). *Learning to See. VSM to Create Value and Eliminate Muda*. Lean Enterprise Institute.

Singh, J., & Singh, H. (2015). Continuous improvement philosophy—Literature review and directions. *Benchmarking: An International Journal*, 22(1), 75–119.

Sinha, N., & Matharu, M. (2019). A comprehensive insight into lean management: Literature review and trends. *Journal of Industrial Engineering and Management*, 12(2), 302–317.

Six Sigma: A Complete Step-by-Step Guide: A Complete Training & Reference Guide for White Belts, Yellow Belts, Green Belts, and Black Belts. (2018). Council for Six Sigma Certification.

Suarez-Barraza, M. F., Ramis-Pujol, J., & Kerbache, L. (2011). Thoughts on Kaizen and its evolution: Three different perspectives and guiding principles. *International Journal of Lean Six Sigma*, 2(4), 288–308.

Voehl, F., Harrington, H. J., Mignosa, C., & Charron, R. (2013). *The Lean Six Sigma Black Belt Handbook: Tools and Methods for Process Acceleration*. CRC Press.

Wickramasinghe, V., & Chathurani, M. N. (2021). Effects of continuous improvement in streamlining HRM practices. *Business Process Management Journal*, 27(3), 883–900.

Womack, J. P., & Jones, D. T. (1996). Lean thinking: Banish waste and create wealth in your corporation. *Harvard Business Review*, September–October, 140–158.

Womack, J. P., Jones, D. T., & Roos, D. (1990). *The Machine That Changed the World*. Simon & Schuster.

3

Co-usage of Improvement Methodologies

3.1 General Concept

Lean Management is widely known among managers as an improvement methodology and is frequently practiced by companies of different types. Also, Six Sigma is very popular, as it offers a well-defined system to implement systematic changes in organizational processes. Lean Six Sigma, as a combination of both, is gaining popularity among companies. At the same time, the Theory of Constraints proposes very attractive steps toward sequential development. The integration and synergistic coexistence of these three methodologies is very important, as this might enhance development and growth by forming a combined, powerful improvement methodology.

The improvement methodology of any kind requires some organizational system to exist in an organization's everyday reality. Each methodology has a different view of how to be practically organized. Lean Management leaves much space here, as this approach covers a wide range of issues and rather integrates with the existing structure, expanding and rebuilding it afterward. In contrast, Six Sigma has multilevel certification of experts assigned to different roles in improvement projects, and project implementation is based on project management methods. The Theory of Constraints, as a kind of general operations management theory, does not define any special organizational structure. All of them offer hundreds of practical methods and tools.

Management principles and assumptions underlying each methodology were discussed earlier in Chapter 2. Each methodology views improvement somewhat differently, but the general aims are still convergent. As the empirical study indicates that the co-use of methodologies is meaningful, many entities use their own improvement methodologies. The Theory of Constraints is one of the methodologies most commonly used alongside others.

The Theory of Constraints implies striving for maximum productivity in one special place of the entire organizational system—the bottleneck. This place is of special importance and attention. However, all other processes should work very efficiently and predictably to ensure the desired maximum flow in the bottleneck. Such a development concept exploits the logic of the theory and leverages the potential of two additional methodologies. There is

DOI: 10.1201/9781003655237-3

no maximum exploitation of the bottleneck without the rest of the processes having low variability and high predictability, and without the successive elimination of waste in all processes, assuring an increase in value and a decrease in costs.

3.2 Introduction

Managers of different types of organizations desire approachable and effective methods and systems to support them in leveraging their process productivity along with the entire organizational system. Popular methodologies offer interesting and attractive paths for organizational improvement, equipped with many specific practical methods and techniques, useful guidance, and best practices.

The methodologies to be implemented within companies must be organized and must have an organizational system to be effective. Each of the three methodologies—Lean Management, Six Sigma, and Theory of Constraints—is generally organized in different ways. They also view how improvements are situated within the whole organizational system slightly differently. Additionally, they may more or less precisely define the roles necessary in the system of regular improvement. They suggest different toolboxes for improvement projects.

Despite the differences, the idea of one integrated improvement system is very tempting. The system which takes advantage of the strengths of each methodology is well-fitted to a perspective of an individual organization and its management philosophy. Such integration might lead to a powerful methodology truly driving the company's operational development. This whole chapter builds on the principles and other characteristics discussed in the previous chapter.

3.3 How Improvement Systems Are Organized

To work effectively, management concepts of any kind need some inner organizational system along with methods for practical use by people in the organization. Thanks to them, the purposes for which they were used can be turned into reality, and managerial efforts become effective.

When it comes to Lean Management, authors first of all underline the fundamental importance of a long-term strategic vision of organizational excellence (Liker, 2004). It should be vivid, clear, strong, and widespread within the organization to stimulate individual initiatives. This long-term vision of organizational development must also be reflected in important breakthrough decisions and everyday managerial decisions. Consequently, the long-term development vision is more important than short-term gains or losses, so sometimes a company has to accept short-term losses to follow a long-term vision.

The long-term vision is the most general and largely fleeting; it can also be explained as the omnipresent will to strive for excellence in the organization. But beyond that, for Lean to function in an organization, Lean policies and Lean practices are indispensably needed (Jedynak, 2015, p. 62). Lean policies, just like the above-mentioned Lean vision and Lean practices, are equally important and necessary. Lean policies are understood as rules and principles that actually function within an organization and determine strategies and real actions toward a certain scope or area of the organization's activities, such as cooperation with suppliers or treatment of employees (McCulloch, 2003).

Lean policies do not constitute empty declarations but are specific rules established by managers in their companies to fulfill long-term intentions. The policies have to be stable for a sufficient period of time to reveal their effectiveness and bring gains to Lean in the organization. If policies are changed quite often, they will not serve as policies in fact. Thanks to them, trust in the long-term excellence vision is built.

Lean policies must be focused on the key issues of Lean Management in the organization. The organization can easily identify them by understanding the Lean approach along with its principles. One of the objectives of Lean policies is obviously the employees. The organization must set rules for its employees so that they trust process improvement changes and face no obstacles in engaging in bottom-up initiatives. For example, the rewards related to improvement projects must be established very judiciously, but fulfilled and maintained over long periods; no one should feel cheated.

The third pillar of tangible Lean existence in the organization is Lean practices. They are a set of techniques that aim to eliminate waste and increase the efficiency of the organization. Lean practices are primarily the practical actions taken by organizations in a specific direction related to Lean assumptions (Wickramasinghe & Wickramasinghe, 2017, p. 533). In fact, the literature in the field, especially managerial manuals, pays most of their attention to Lean tools, which are usually the flagship Lean practices in companies. Some authors report over 100 Lean tools being used by organizations (Pavnaskar et al., 2003); however, when considering their multiplied versions and frequent adaptations from related areas, there is no point in enumerating all of them.

Some Lean tools play a special role in the Lean Management organizational system; some are even perceived as Lean symbols. One such tool, which directly serves to organize the workspace, is 5S. The 5S method is universal and can be implemented in various workplaces, from manufacturing to office environments. The method consists of five steps, which are applied and then focused on maintaining the established solutions, with the steps repeated periodically. The steps of 5S are selection, systematization, cleaning, standardization, and self-discipline (Chandrayan et al., 2019). The method ensures improved workplace safety, increased productivity, better product quality, enhanced employee morale, and a greater sense of responsibility for the work performed.

As 5S is considered by many a must-have starting tool, value stream mapping (VSM) is not as popular; however, it is a fundamental analytical and planning method that touches the heart of the Lean concept. VSM is a practical solution for describing, diagnosing, and designing the product flow throughout the entire company in which value for customers is produced. VSM analyzes, using a visual approach, the flow of materials and steering information from supplied material to product delivery to customers (Chen & Meng, 2010). The mapping consists of a stage of diagnosing the current value stream, which includes identification and estimation of existing waste, and a stage of designing an improved stream, which is significantly better in fundamental characteristics and contains less waste.

These two Lean tool examples provide a simplified overview of their possible roles in the Lean system. When considering the typical functions of all Lean tools, they cover the following purposes:

- workplace organization along with keeping machines in permanent readiness for work,
- product flow and workflow coordination,
- quality of work and products improvement,
- problem analysis, in particular discovering their causes,
- set directions of improvements,
- support directly performing improvement projects, along with teamworking.

However, it must be emphasized that although Lean tools are inevitably a fundamental part of Lean Management's practical functioning, they do not encompass all Lean practices. Apart from standardized, widely elaborated Lean tools described in sources, there exist many individually designed and tailored Lean practices to organizational needs, such as routines, means of work standardization, ways of information exchange, and forms. In fact, they largely contribute to the efficient functioning of the entire Lean organizational system.

At the same time, it must be firmly underlined that Lean Management cannot be identified with Lean tools. Many studies emphasize that the key to Lean Management's success lies in going beyond implementing tools and thoroughly understanding the philosophy behind this approach, which seeks out waste and gradually eliminates it over the long term (Liker, 2005). The author also warns against "implementing Lean using PowerPoint presentations" and "cherry picking" strategy. This means preparing numerous presentations but not putting Lean into action, and arbitrarily selecting only the easiest Lean tools to implement, without addressing the whole crucial components of this approach, including the most challenging ones. These will eventually cause Lean to fail.

Lean Management is a system that permeates the entire organization; therefore, it obviously integrates with the entire organizational structure in order to adapt it accordingly to the logic and principles of this approach. However, according to some authors, its introduction to organizations requires building a Lean team (Karim & Zaman, 2013). The team should consist of experts and management staff from various departments. The task of this team is to organize people and resources to implement Lean. It should be noted that there is a lot of space and diversity in how the structure directs responsibility for the Lean system's maintenance.

In terms of the structuration of the system, the Six Sigma approach is noticeably more specific and formalized. However, the fundamental issue is how systematic changes in processes are practically performed. While carrying out improvements in processes, several kinds of tools and techniques are employed. First of all, statistical techniques are widely used in improvement projects. It is characteristic that numerous statistical techniques are systematized according to the level of advancement, which is closely related to the concept of the Six Sigma system in the organization.

Authors, for example, classify them into three categories—basic Six Sigma tools, intermediate tools, and advanced tools (Sharma & Chetiya, 2009, p. 14). Below is an example list of statistical knowledge recognized as the basic statistical toolkit in Six Sigma (Voehl et al., 2013, p. 557):

- types of scales, types of variables, data acquisition, observations, sampling, and determining data precision,
- binomial, normal, and Poisson distributions,
- descriptive statistics: mean, median, mode,
- histograms, time charts, control charts,
- process capability testing and analysis, upper and lower control limits, process capability indices Cp and Cpk,
- Chi-square test, correlation coefficients,
- variance and analysis of variance (ANOVA),
- statistical hypothesis testing.

The above overview of Six Sigma statistical techniques at the basic level accurately illustrates the characteristics of the Six Sigma approach to performing changes in processes. The toolkit contains not only various statistical tools but also a set of research techniques, including data gathering. Generally, it is clear that practicing Six Sigma, even at the basic level, undoubtedly requires fairly advanced statistical knowledge and skills. The next levels of Six Sigma tools differ from the basic level primarily in the use of more advanced and complex statistical tools.

To clarify the scope of the basic level of the Six Sigma toolkit, apart from the mentioned statistical techniques, there are also plenty of other improvement techniques in practice. Most of them are directly adopted from the Total Quality Management approach. There are seven widely known problem-solving tools: process flowcharts, cause-and-effect diagrams, data collection checklists, process monitoring checklists, and data visualization tools. Another set of tools, named 7M, was developed by the Union of Japanese Scientists and Engineers to support innovation and the effective implementation of improvement projects (Tague, 2024). The 7M set includes the relationship diagram, interdependence diagram, tree diagram, matrix diagram, prioritization matrix, arrow diagram, and decision diagram.

Thus, the Six Sigma system, in its broadest utilization by employees, involves using a wide variety of tools, including analytical and quite advanced statistical techniques, data gathering methods, and many methods supporting the execution of change projects. The system is entirely based on tools intended not only to support particular types of actions but also to guide participants through all stages and respective challenges of the Six Sigma improvement routine DMAIC and its design version DMADV. However, the tools are not assigned to particular steps, and many of them are recommended for multiple uses.

DMAIC, as explained in the previous chapter, is a sequence of steps guiding process change. It is a flexible and universal model, applicable to any type of process and any business sector. Each phase relies heavily on data collection and analysis, ensuring that decisions are based on factual information rather than assumptions (Six Sigma, 2018, p. 158). The DMAIC sequence is applied to solving any issues in processes, that is, related to quality, cost savings, improving capability, and also improving customer satisfaction.

The Six Sigma system in an organization is managed on two levels. The first level concerns the entire organization and is performed by senior managers, while the second is the team level. Senior leaders set overall goals, approve projects, ensure adequate resources, and evaluate results. Senior leaders are responsible for promoting the Six Sigma culture, aligning project goals with business objectives, and rewarding performance. At the team level, the focus is on building effective teams, defining roles, and managing schedules, budgets, and success metrics for each project (Six Sigma, 2018, p. 142).

Several organizational roles are defined in the Six Sigma system. The functioning of the Six Sigma system is based on these roles, and its proficiency depends on them. These roles are as follows (Pyzdek & Keller, 2010):

- Senior Leaders and system Steering Committees,
- Black Belts, Project Leaders,
- Green Belts,
- Process Owners,
- team members.

The top management in the Six Sigma system points out people who support and guide improvement projects and promote the idea of process improvement from their level through the entire organization. Therefore, roles such as Senior Leader, along with a steering board and a certification board, are established.

Black Belts and Green Belts play special roles in the system. Employees acting as Black Belts serve as change project leaders. They have broad and comprehensive knowledge of Six Sigma methods and tools, especially statistical techniques, and they also have extensive technical knowledge. They lead project teams through the DMAIC process, and their main tasks include training team members, managing the project from start to finish, and ensuring that the project meets its objectives, leading to real improvements (Pyzdek & Keller, 2010, p. 20).

Green Belts are team members who have undergone Six Sigma methodology training, but to a lesser extent than Black Belts. They typically work on smaller projects as leaders or assist Black Belts. Their role is primarily to support project teams, collect data, and assist in implementing process improvements. It should be noted that in the Six Sigma system, there are more certified levels of initiation; there are also Master Black Belts, Yellow Belts, and White Belts. The latter represents the basic level of knowledge of the methodology, meaning familiarity with basic concepts and fundamental thinking about quality and processes, as abbreviated above.

In the Six Sigma system, members of project teams are people from different departments who contribute their expertise to projects. They actively participate in identifying problems and searching for solutions, playing an active role in many solution-seeking techniques. The implementation of the agreed changes relies on members of project teams. They ensure a variety of perspectives on problems and possible solutions. Special members, who are experts in selected fields such as advanced data processing tools, are appointed to project teams.

In organizations practicing the Six Sigma approach, it is common to assign owners to organizational processes. Process Owners have cross-functional knowledge of the processes they manage, as well as the challenges and difficulties that arise in those processes. The task of Process Owners is to work closely with project teams to ensure that improvements are effectively integrated into day-to-day operations and that the changes made are sustainable.

The application of Six Sigma requires numerous new roles within the organization; representatives from all levels of the organizational structure have to be actively involved. Active support and personal involvement of board members are necessary to create favorable conditions for implementing changes in the organization's processes. Top management should promote the vision of six sigma-level process excellence, and the organization should be structured so that all employees treat the pursuit of process excellence with minimal variability as their normal daily duty (Pyzdek & Keller, 2010, p. 13).

The literature also provides some insights into how the Lean Six Sigma system is organized. According to Taghizadegan (2006), the Lean Six Sigma system uses structures analogous to Six Sigma. At the head of the Lean Six Sigma system are representatives of top management who act as system leaders for the entire organization or for a business unit (Corporate Champions). System leaders, having the authority of a management board or taking part in a management board, and also having responsibility or co-responsibility for the company's results, are responsible in the Lean Six Sigma system for translating strategic goals into improvement projects focused on creating value in the organization.

Improvement work is organized in project teams, with roles ranging from Black Belts, who are project leaders, to Green Belts, who participate in the teams and provide support to leaders and participants. The Lean Six Sigma system includes a hierarchy of roles that supports the integration of Lean Six Sigma projects into the organization's daily operations. The goal is to build a Lean Six Sigma system that enables effective project management, integrated activities, and a focus on financial results, which should be converted into increased shareholder value (Taghizadegan, 2006, p. 23).

The Theory of Constraints is much less equipped in the executive layer specific to this approach. At the heart of the theory lies the Five Focusing Steps logic. According to it, it is essential to focus solely on a single, selected issue related to a constraint—that is, a factor that currently limits the organization's growth (Goldratt, 1984). The Five Focusing Steps is not definitely an improvement procedure such as DMAIC, for example. It is rather a managerial logic that helps to guide changes from an organization-wide perspective.

The analytical tool designed to be used in this approach is Throughput Accounting. This tool is used to calculate what is more beneficial and what is less beneficial for a company; it refers to volumes of products or possible changes in processes. It is an original calculation model developed as an analytical component supporting this approach. The throughput calculation is designed to avoid the weaknesses of traditional cost accounting, which provides limited support and may even hinder managerial decisions (Corbett, 1998).

The Theory of Constraints, supported by Throughput Accounting, allows for maximizing not only production volumes but also the profits generated by the company. While many companies prioritize reducing operating costs, throughput calculation primarily leads to increasing the company's throughput. Taking throughput calculation into consideration helps avoid making operational decisions based on the local conditions of individual processes or single indicators, as it forces a holistic approach to the organizational system.

The Theory of Constraints proposes a practical tool for maintaining an equal pace of flow subordinated to the rhythm of the system bottleneck. The DBR (Drum, Buffer, Rope) method has a planning and control mechanism that meets the requirements of the Theory of Constraints. It allows for an even flow of products through the system at the maximum rate, which is determined by the full utilization of the bottleneck's capacity. According to the DBR method, the slowest process in the system or the market determines the pace of all other processes; otherwise, the inventory in the system will grow uncontrollably (Cox & Schleier, 2010, p. 146).

DBR introduces bottleneck schedules that drive the flow of products through the entire system. At DBR, a work scheduling system is conscientiously prepared for the bottleneck, which is symbolically referred to as a drum. The production schedule for the bottleneck process must consider the maximum flow, that is, the maximum utilization of the process's capacity. The safety stock is also defined; this is a buffer. Finally, the rope triggers the flow to the bottleneck from the previous processes.

The literature does not suggest introducing a special organizational system for the Theory of Constraints. This seems understandable, as it is a general concept covering the entire operational management of the organization. There are also limited guidelines supported by grounded managerial experience on how to apply the methods mentioned above.

3.4 Challenges and Shortcomings

It is absolutely obvious that process improvement methodologies have specific strengths but also limitations. There are still plenty of challenges in front of the methodologies as they have to serve effectively for companies in organizational improvement struggles. The issue is not often a methodology itself but rather its implementation in the company.

Scholars note the possibility of negative effects, mainly social ones, if Lean Management is implemented incorrectly in organizations. These effects may include reduced job satisfaction resulting from a loss of autonomy and lack of employee engagement, excessive working hours and a feeling of low well-being, or increased stress at work. Such effects, which are in fact the opposite of what Lean is all about, can occur when the methodology is misunderstood, resulting in fundamental errors during its implementation (Salentijn et al., 2021).

One of another possible mechanisms is a pressure to achieve quick results in terms of improving process indicators without considering the mechanisms and methods used to achieve them in Lean Management. A pressure to achieve high level of some indicators, which are totally incomprehensible and unacceptable to the staff and perceived as the execution of arbitrary interests of the management, might also happen. These are also undoubtedly misunderstanding of the Lean philosophy.

The appropriate approach to the people in the organization during the implementation and further development of Lean is primary important. For example, a study noticed that the traditional piecework-based remuneration system had a negative impact on Lean in a company. Employees were unwilling to stop production in accordance with the Jidoka principle, for example, because it affected their remuneration (Robertsone et al., 2022). Moreover, the study concludes that changing the remuneration system alone, without a holistic approach and substantial change in the company's values, does not bring full effect.

Some literature voices are trying to identify and systematize the challenges that organizations often face when they undertake to implement and follow the Lean approach. Based on the Finnish furniture and shipbuilding industries, a summary of challenges is as follows:

- A short-term orientation may damage Lean Management in the organization. Companies might focus on immediate profits and ignore the long-term benefits of implementing Lean. The lack of a long-term vision along with resistance to change make it difficult to follow Lean principles.
- Insufficient diffuse of organizational culture of learning among the entire staff. The success of Lean inevitably requires building a learning culture where employees are committed to process improvement and permanent reflection.
- Balancing the flow of work throughout the system. According to surveyed companies, organizing the flow according to the principles of balanced flow is a major difficulty.
- Long-term relationships with suppliers along with Just-in-Time cooperation. Many companies preferred short-term gains at the cost of stable relationships with suppliers.
- And finally, achieving a sense of empowerment among employees. Employees are limited to scope of their tasks with no space for empowerment and no opportunity to propose improvements (Rymaszewska, 2014, p. 1000).

Similar challenges emerge from other studies among Latvian light industry companies (Robertsone et al., 2022). Researchers identified a lack of thorough understanding of the Lean philosophy as the main problem. In addition, the study indicated strong resistance among employees, which may be related to poor commitment and lack of leadership from top management, as well as poor communication within plants. Other difficulties closely correspond to the challenges listed above.

There are also identified challenges referred to Six Sigma that companies face when following this methodology. The Six Sigma system requires establishing new positions and departments, and the incentive system must be adjusted to include rewards, recognition, incentives, and compensation related to this system. The knowledge and skills required for the basic roles in the system, from White Belt to Master Black Belt, require significant investment in employee development, as well as time-consuming and effective training to ensure that these employees can correctly perform process improvement projects (Schmidt et al., 2018).

There are several factors that contribute to Six Sigma's success. These include:

- Accurate prioritization and selection of improvement projects in the organization.
- Project management skills—high competence in this area is key to the success of the process improvement being implemented.
- Linking Six Sigma system to business strategy—consistency with the organization's goals increases the value of improvements.
- Effective internal communication along with communication to stakeholders are both important to avoid barriers and resistance to change.
- Connecting Six Sigma with customers—a deeper understanding of customer needs is at the heart of improvement projects, and customers should be involved in change projects as appropriate.
- Accurate reporting and consideration of qualitative data—if changes are based on solid data and transparent reporting, effective progress monitoring is possible.
- Linking Six Sigma to suppliers—cooperation with suppliers enables significantly better results in improvement projects and greater support for the organization's goals (Singh & Singh, 2020, p. 795).

The Six Sigma system is a fairly structured and formalized approach to continuous process improvement in an organization. It relies heavily on analytical knowledge and complex statistical tools and models. Implementing changes in processes involves a series of change in projects carried out by well-prepared teams of experts and rest of the employees within the organization. In fact, it also requires the engagement of employees in improvement projects.

There are also experiences in terms of integrated Lean Management and Six Sigma use by organizations. Authors try to specify several difficulties when integrating two methodologies into Lean Six Sigma; they are as follows:

- First of all, there is a noticeable reluctance among Lean practitioners to use statistical tools, which are in fact pretty advance and at the same time a key to Six Sigma.
- Typically, Lean practitioners are focusing primarily on quick, visible results, while Six Sigma prefers more analysis and a longer

implementation time; this leads to a certain contradiction that makes it difficult to integrate both approaches.

- There is a clear difference in organizational culture and general approach to process improvement in both methodologies. So, improvements are expected to be more flexible and reactive in one approach and more top-down planned and formalized in the second.
- In the area of training and knowledge acquisition, there may also be a barrier to the harmonious application of both approaches. Six Sigma requires wide statistical knowledge, which may be completely incomprehensible to people practicing Lean (Rodgers et al., 2019, pp. 644–646).

Among guidelines for Lean Six Sigma implementation, authors also emphasize the importance of focusing on the simplicity of the tools and techniques implemented in the organization, avoiding the use of complex statistical tools, especially in the initial phase of Lean Six Sigma implementation. A well-thought-out and clear roadmap-style action plan is needed, it has to determine the order in which Lean Six Sigma tools are implemented (Albliwi et al., 2015, p. 679). The lack of such guidelines will hinder the implementation of this approach, especially in its early stages.

Moving on to the Theory of Constraints it seems that there are much less studies on limitations and challenges in front of implementation of this methodology. At the same time authors underline its great potential for practical use in manufacturing companies along with universality (Datt et al., 2024; Lewandowska-Ciszek, 2018). Moreover, the Theory of Constraints is the only methodology which proposes a dedicated approach to calculating profitability that would support decisions related to process improvement—Throughput Accounting.

Despite this methodology offering a number of practical studies and original methods, it also has areas that are insufficiently developed. For example, there are methods available for identifying production bottlenecks, but in the case of complex production flows, there are still significant doubts as to the accuracy of the system bottleneck (Urban, 2019). The application of the Theory of Constraints poses difficulties when there are so-called floating bottlenecks, that is, bottlenecks that occur for short periods of time, and then appearing in another part of the system. There are also no detailed guidelines for identifying and improving constraints outside the organization's system.

Differences between methods, along with their strengths and limitations, would be factors leading companies to use more than one methodology. However, this may also be due to a desire to find a more appropriate methodology when one has encountered some difficulties or yielded unsatisfactory results. The joint use of multiple methodologies will be discussed next in this chapter.

3.5 Use and Co-use of Methodologies—Empirical Case Study

Many methodologies of process improvement are available; however, three of them are of special importance in terms of regularly effective changes in processes. It is interesting to consider how frequently these methodologies are practiced by companies, and whether they are used separately or a few of them at the same time. Reliable answers to such questions allow for a better understanding of the methodologies themselves and their utilization by organizations.

The use of improvement methodologies has been investigated on a sample of 380 randomly chosen medium and large companies operating in Poland. The sample is statistically representative of these size categories in the country. The survey, as presented in the previous chapter (Chapter 2), included questions regarding the use of process improvement methodologies. Besides the aforementioned methodologies, the option of a tailored methodology prepared by a company was also taken into consideration, as this is a real option, especially in large companies and international corporations. The survey encompasses the following methodologies, with abbreviations included:

- Lean Management, LM
- Six Sigma, SS
- Theory of Constraints, TC
- Tailored Improvement Methodology, TIM

In Table 3.1, there are data on the declared use by investigated companies on the improvement methodologies. Among methodically separated systems, Lean Management is the most frequently practiced in investigated sample, almost 19% of companies declared to follow Lean methodology, Six Sigma methodology only slightly less, almost 16% of the surveyed entities. The Theory of Constraints was declared to be followed by 9.5% of companies. Finally, individually developed methodologies were declared to be the most often used, almost 37% of companies declared to use such individually tailored own methodology. The significantly high use of own tailored

TABLE 3.1

Use of improvement methodologies in the sample

Methodology	Number	Percentage
LM	72	18.9%
SS	60	15.8%
TC	36	9.5%
TIM	140	36.8%

TABLE 3.2

Test for differences in the use of methodologies

Chi-square	df	*p*
Prevailing activity type and LM		
12.504	3	0.00584
Prevailing activity type and SS		
9.038	3	0.02879

TABLE 3.3

The use of methodologies by companies in activity types

Prevailing activity type	LM		SS		No. of comp.
	No.	Per.	No.	Per.	
B2B services	**18**	**29.5%**	12	19.7%	61
Consumer services	18	12.5%	**28**	**19.4%**	144
Manufacturing	**23**	**26.4%**	15	17.2%	87
Other types of activities	13	14.8%	5	5.7%	88
Jointly	72		60		380

approaches may indicate a strong need for better improvement methodologies meeting the needs of companies.

The survey also identified the types of prevailing activities of the respondent's companies. Thus, it was possible to identify if the use of improvement methodologies differs between business activities. To check whether there are differences between types of activities, the Chi-square statistical test was performed. The output is presented in Table 3.2. The use of the two methodologies, Lean and Six Sigma, is statistically different in particular types of activities, where p-value is less than 0.05. Information on the use of these methodologies in particular types of activities is presented in Table 3.3.

The data referred to the use of two methodologies, in particular in business sectors (Table 3.3), show that Six Sigma is the most popular in customer services. 19.4% of companies operating in this sector use Six Sigma, while only 12.5% of them practice Lean Management, which generally is slightly more popular. On the other hand, in manufacturing as many as 26.4% of companies use Lean Manufacturing while only 17.2% use Six Sigma. Also, in business-too-business (B2B) services there is a noticeable difference in use of these methodologies—Lean Management is used by 29.5% companies and Six Sigma by 19.7%. So, while Six Sigma is especially used by customer service companies, Lean Management above all by manufacturing and B2B services companies.

Gathered data allow to investigate if the methodologies are co-used by companies. In Table 3.4, the numbers and percentage of respondents who declared one methodology and at the same time another one are presented. The table has to be read in lines, for example, 17 companies which use Lean

TABLE 3.4

Co-use of improvement methodologies

	LM		SS		TC		TIM	
	No.	Per.	No.	Per.	No.	Per.	No.	Per.
LM	72	100%	17	23.6%	12	16.7%	18	25.0%
SS	17	28.3%	60	100%	9	15.0%	12	20.0%
TC	**12**	**33.3%**	**9**	**25.0%**	**36**	**100%**	**13**	**36.1%**
TIM	**18**	**12.9%**	12	8.6%	13	9.3%	140	100%

Management also declared using Six Sigma, which is 23.6% of companies practicing Lean, fourth and fifth column, first line below headlines (Table 3.4).

From the data emerges the view that companies quite often use more than one improvement methodology; the minimum percentage is 8.6% and the maximum is 36.1% for using one methodology while also using another (Table 3.4). The data allow us to observe some more characteristic behaviors of companies. Among companies that decided to build their own tailored methodology, there are also companies using other standardized methodologies. As might be expected, there are not many of them; however, almost 13% of those who declared having an individually tailored methodology also declared using the Lean Management approach. Thus, an individually designed methodology might be practically supported by other standardized methodologies.

However, the most important observation emerging from these data is the usage of the Theory of Constraints. This less frequently practiced methodology is characterized by the highest rate of co-usage among the four approaches examined in the investigation. For example, as many as 33.3% of companies declaring use of the Theory of Constraints also declared usage of Lean Management, and up to 36.1% of them declared that they have an individually designed methodology for process improvement (TIM).

The data suggest that the Theory of Constraints is a type of horizontal methodology often employed when another approach is already implemented. This implies a special role for the Theory of Constraints in improvement systems built by companies, including both tailored and standardized approaches. When discussing the co-usage of improvement methodologies in companies, it is not only important that multiple approaches coexist, but also how to use several methodologies effectively to obtain additional benefits and synergies.

3.6 Toward Constraint-Driven Lean Six Sigma

As proved in the section above, the coexistence of different methodologies is already quite widespread among the investigated companies. The issue of coexistence and integration of multiple approaches within one organizational system appears to be quite important for managers right now.

However, some interesting thoughts on the joint use of the three methodologies have already been revealed. For example, the idea of the Velocity concept (Dee et al., 2009) is to achieve high performance by combining the strengths of the three methodologies, thus eliminating waste, reducing variation, and removing system bottlenecks.

As the Theory of Constraints offers a quite complex operational theory of the organization along with practical thinking and several routines that help improve processes, it nonetheless assumes high predictability and stability of the organizational system's flows. It also views the organization as almost as algorithmic as mechanical systems are. These aspects may appear very attractive to managers but are not necessarily fully feasible in an exact manner due to the complexity, variability, and unpredictability of the internal and external organizational environment.

At the same time, Lean Management and Six Sigma, which are now often practiced together as Lean Six Sigma, can boast impressive results, especially in some flagship companies. However, many companies also report disappointment with the results achieved by introducing these improvement methodologies. The methodologies are also criticized for being suitable only for mass production systems and for requiring people with PhDs in statistics.

Moreover, the lack of integration of process improvement into the organizational structure and management system prevents improvements from developing and producing the desired results (Keim, 2011). The problem may also be a lack of systemic thinking in the organization that would incorporate process improvement and integrate it with existing organizational practices (Breyfogle, 2005). Changes in processes might be problematic if improvement projects are not consistent with the company's strategy, are inconsistent with business objectives, or are not sufficiently focused on benefits that are vital to the company (Kornfeld & Kara, 2013; Snee, 2010).

This is a utopia in which each of these methodologies can be exploited freely, taking some pieces from one methodology and others from different ones, without having a coherent concept of how to use them together and achieve a strong system of process improvement. Each of these approaches has all the elements that make up its whole. In order to create a new quality, it is necessary to skillfully select the logic, assumptions, principles, methods, and organizational system to create a new methodology that offers more than each of the three used separately.

Constraint-Driven Lean Six Sigma framework is necessary to exploit the strengths and particularly useful functionalities of each methodology, while simultaneously eliminating possible contradictions. It strives to build a powerful system for improving processes in organizations that meets the expectations of modern organizations and managers. Such a framework must be based on a deep understanding of these three methodologies.

Each of these methodologies contributes great and original ways of thinking, principles, and sets of useful methods that can serve as material for building new organizational quality and an effective system for improving processes. Lean Management focuses on eliminating waste and improving the flow of processes throughout the organization. Six Sigma contributes a data-driven approach for the permanent reduction of variations in processes. Finally, the Theory of Constraints provides a systemic approach for focusing efforts on the weakest organizational points to achieve the greatest impact on the organization's results.

The framework of Constraint-Driven Lean Six Sigma must consider crucial organizational spheres and issues indispensable for the existence of an improvement system. One of the fundamental issues in an effective process improvement system is the development vision determined by managers and disseminated among people.

Striving for excellence is a foreground motive in the whole Lean approach; it forms the cornerstone of an organizational culture of excellence. However, the vision of excellence in processes is possibly one of the hardest things for Western managers, as it might seem to them—business people with their feet firmly on the ground—as too abstract and too idealistic. The general theory of operations management comes to the rescue with its concept of organizational constraints.

The Theory of Constraints allows the formulation of the operational strategy for the company. This goes hand in hand with the development vision of the company. According to the theory, the organization should strive for very specific things—one place in the system which is crucial (the bottleneck)—and all other processes must work very efficiently and predictably to ensure maximum flow through this bottleneck. This is a very specific and tangible vision: moving from one bottleneck to the next to improve them significantly and continuously enhance the operational efficiency of all other related processes.

Such a vision is much easier for managers to accept and to propagate within the organization. However, it cannot be left without a component of systematic process improvement toward excellence in all processes, in terms of their variability and elimination of waste. It should be emphasized that only stable and highly predictable processes, with low variability, for supplying the bottleneck and picking up from it, can assure maximization of the flow through it. Furthermore, only the elimination of non-value-adding activities can assure gradual operating cost rationalization. The fundamental assumptions of Lean Management and Six Sigma can take on a slightly new meaning in the light of the Theory of Constraints.

The organizational culture of improvement is an indispensable issue when it comes to process improvement; this is underlined by all three concepts. However, it is Lean Management that contributes the most to the field of organizational improvement culture (continuous process improvement). The Kaizen culture, which is the backbone of Lean, is an organizational culture of small and frequent positive changes performed by a wide range of employees.

The organizational culture of employee engagement allows the organization to achieve ever-decreasing variability and greater predictability and controllability of processes, which is absolutely essential. Without mass improvements—small but frequent and omnipresent—high efficiency and low costs in processes cannot be achieved. Only massively present changes, using the knowledge and creativity of regular operators and their involvement in implementing the changes, lead to high-performance processes.

It should also be noted that, in many cases, striving for variability reduction in processes goes hand in hand with the elimination of waste, if the former is correctly understood. Both seem to be mirror images of each other. However, the real headache is the "islands of excellence" in noncritical areas, which possibly occur when following only Lean Six Sigma (Nave, 2002; Moura, 2012), especially when these islands of excellence cost the organization much.

The contribution of the Theory of Constraints comes to the aid of this issue with a very clear and attractive concept. Now, this is not only a general vision but also specific tools for determining what exactly should be improved in the organizational system to maximize throughput and, as a consequence, financial benefits. The Theory of Constraints approach allows identification of bottlenecks and other constraints along the flow and demonstrates suggestively how much they need to be improved and what level of benefits this will bring to the organization.

The Theory of Constraints, with its methods of bottleneck identification, allows us to determine which machine or process needs to be elevated and how much its throughput should be improved, expressed in figures. Then, the Theory of Constraints, along with its Five Focusing Steps (5FS), contributes to the prioritization of improvement ideas and projects. Its logic implies improving the bottleneck first, followed by the maximum simplification and elimination of all waste in all processes supplying the bottleneck and from the bottleneck to the customer's delivery.

For example, the Theory of Constraints concept can be combined well with the VSM method of Lean Management. VSM is a handy tool for mapping all the processes working together in a stream, from door to door. VSM models the value stream in a visual manner and is used for designing improvements in the stream. The improvements could be designed according to the Theory of Constraints logic. A bottleneck in the stream must be identified, maximally elevated, and all the other processes subordinated to the planned rhythm of the bottleneck.

Combining VSM with the Theory of Constraints can supplement fundamental Lean analytical techniques with valuable tips, making it much more useful and widely practiced by managers and employees. Formulated improvement plans will be more accurate and will approach the best possible ratio of improvement efforts to achieved results.

Lean Management is focused not only on the internal company stream but also on all links of the cooperation value chain, as clearly advocated by Lean Thinking (Womack & Jones, 2003). The theory's search for bottlenecks should also focus on bottlenecks at suppliers and recipients. The perspective of the whole value chain is not sufficiently emphasized by the Theory of Constraints literature, but the Lean approach highlights the necessity of this perspective.

Lean Management is often explained by management principles underpinning this improvement approach. The fundamental principles of Lean and Six Sigma are still valid, focusing on the value stream, producing value, eliminating waste, and emphasizing the customer, which are the primary motives of both Lean and Six Sigma, as well as reducing variability to improve predictability.

To unleash its full potential, the organization requires many types of improvements of various sizes. Large-scale, groundbreaking improvements that require time, greater effort, and resources provide significant development. However, small but numerous improvements delivered over time also bring impressive results in terms of process indicators such as Lean Time, unit costs, and quality. Both kinds of improvement should exist in the organization, and tailored methodologies should be assigned to each.

Small bottom-up initiatives can employ selected routines from various change methods offered by Lean Management. The Kaizen routine can be tailored to the organization's business profile and engagement culture. For greater improvements, Six Sigma's routines might work better. Considered methodologies propose many organizational routines for carrying out improvements. The organization should choose the most suitable improvement routines, ensuring efficient implementation of changes and utilizing the potential of engaged employees.

Implemented improvement routines are highly interdependent with the organizational system responsible for improvements within the organization. This means that complex and formalized structures are less conducive to bottom-up initiatives, and the improvement routine must be more formalized. However, focusing on bottom-up improvement initiatives requires that routines be adapted accordingly.

The Six Sigma system structure, being very well developed, may give the impression of being the finest model to adopt in an organization following the combined methodology of Lean Management, Six Sigma, and the Theory of Constraints. However, a highly formalized improvement organizational system, along with improvement projects, may not necessarily be conducive to maintaining and stimulating people's engagement. Lean Management indicates a structure much more embedded in the functions and operations of the organization.

If the organization wants to enable and support employee initiatives and engagement, the system of improvements has to be more open, less formalized, and more decentralized. The organization should shift from the strict logic of top management–accepted projects toward support and advice for bottom-up initiatives.

According to proven knowledge, one factor is absolutely indispensable: the improvement system must consider top management involvement. Actions taken, whether small or large, should have the full support of top management. The organizational improvement system must ensure a good link between managers and improvement projects; all support for improvement projects must be delivered effectively.

Knowledge about implementing changes and basic methods and techniques, many of which are available in the three methodologies, should be provided according to the scopes in which employees can potentially be involved. Any method or technique from a wide range of possibilities that can be useful should be adapted by improvement projects.

The numerous statistical techniques used by Six Sigma require special attention. They should not be applied automatically, but their potential should be exploited wherever they bring extra value to improvement projects. Today's organizations are digitalized and data-based; hence, improvement projects often need professional support—not only from Six Sigma belts but mostly from data engineers at different stages of improvement.

As the statistical knowledge typical for Black and White Belts (Six Sigma system) might be supportive in some improvement projects, in the era of artificial intelligence (AI) agents, the use of a variety of AI-based tools in projects is much more valuable. Much more important than learning a wide range of statistical techniques is preparing simple AI-based tools that can be used by selected team members in improvement projects.

The cloud of key issues for the integrated Lean Management, Six Sigma, and Theory of Constraints improvement methodology is presented in Figure 3.1.

The integrated Constraint-Driven Lean Six Sigma approach implies that it is not enough to improve just one point in the organizational system, as a

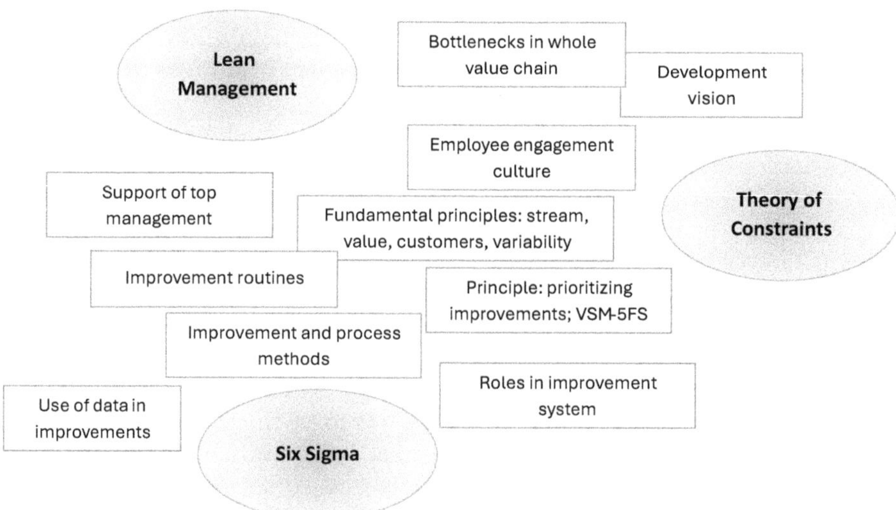

FIGURE 3.1
The key issues in Constraint-Driven Lean Six Sigma.

superficial understanding of the Theory of Constraints might suggest. The organizational system indispensably needs variability in processes to be reduced, as well as all kinds of waste systematically eliminated. Only these measures assure that all processes will work predictably and productively in the rhythm of the system's bottleneck.

References

Albliwi, S. A., Antony, J., & Lim, S. A. H. (2015). A systematic review of Lean Six Sigma for the manufacturing industry. *Business Process Management Journal*, 21(3), 665–691.

Breyfogle, F. W. III. (2005). 21 common problems (and what to do about them). *ASQ Six Sigma Forum Magazine*, 4(4), 35–37.

Chandrayan, B., Solanki, A. K., & Sharma, R. (2019). Study of 5S lean technique: A review paper. *International Journal of Productivity and Quality Management*, 26(4), 469–491.

Chen, L., & Meng, B. (2010). The application of value stream mapping based lean production system. *International Journal of Business and Management*, 5(6), 203.

Corbett, T. (1998). *Throughput Accounting: TOC's Management Accounting System*. North River Press.

Cox, J. F., & Schleier, J. (2010). *Theory of Constraints Handbook*. McGraw-Hill Education Ltd.

Datt, M., Gupta, A., Misra, S. K., & Gupta, M. (2024). Theory of constraints in healthcare: A systematic literature review. *International Journal of Quality & Reliability Management*, 41(6), 1417–1456.

Dee, J., Bergland, S., & Cox, J. (2009). *Velocity: Combining Lean, Six Sigma and the Theory of Constraints to Achieve Breakthrough Performance*. Free Press.

Goldratt, E. M. (1984). *The Goal*. North River Press.

Jedynak, P. (2015). Lean management implementation: Determinant factors and experience. *Jagiellonian Journal of Management*, 1(1), 51–64.

Karim, A., & Zaman, K. A. (2013). A methodology for effective implementation of lean strategies and its performance evaluation in manufacturing organizations. *Business Process Management Journal*, 19(1), 169–196.

Keim, E. M. (2011). Shortcuts don't get you there faster. *ASQ Six Sigma Forum Magazine*, 10(3), 31.

Kornfeld, B., & Kara, S. (2013). Selection of Lean and Six Sigma projects in industry. *International Journal of Lean Six Sigma*, 4(1), 4–16.

Lewandowska-Ciszek, A. (2018). Theory of constraints as a stimulus towards warehouse transformation process on the example of the distribution center. *Management and Production Engineering Review*, 9(4), 96–105.

Liker, J. K. (2004). *The Toyota Way: 14 Management Principles from the World's Greatest Manufacturer*. McGraw-Hill.

Liker, J. K. (2005). *The Toyota Way Fieldbook: A Practical Guide to Implementing Toyota's 4Ps and the A3 Management Process*. McGraw-Hill Companies.

McCulloch, W. M. G. (2003). *Lean Implementation: A Complete Execution Manual for Any Size Organization*. J. Ross Publishing.
Moura, E. C. (2012, June 6). *Using TOC-TP as a Guide to Integrate Lean, Six Sigma, Process Management and TOC Solutions for Business Improvement* [Conference presentation]. TOCICO 2012 Conference.
Nave, D. (2002). How to compare Six Sigma, lean and the theory of constraints. *Quality Progress*, 35(3), 73–78.
Pavnaskar, S. J., Gershenson, J. K., & Jambekar, A. B. (2003). Classification scheme for lean manufacturing tools. *International Journal of Production Research*, 41(13), 3075–3090.
Pyzdek, T., & Keller, P. A. (2010). *The Six Sigma Handbook: A Complete Guide for Green Belts, Black Belts, and Managers at All Levels* (3rd ed.). McGraw-Hill.
Robertson, G., Mezinska, I., & Lapina, I. (2022). Barriers for Lean implementation in the textile industry. *International Journal of Lean Six Sigma*, 13(3), 648–670.
Rodgers, B. A., Antony, J., He, Z., Cudney, E. A., & Laux, C. (2019). A directed content analysis of viewpoints on the changing patterns of Lean Six Sigma research. *The TQM Journal*, 31(4), 641–654.
Rymaszewska, A. D. (2014). The challenges of lean manufacturing implementation in SMEs. *Benchmarking: An International Journal*, 21(6), 987–1002.
Salentijn, W., Beijer, S., & Antony, J. (2021). Exploring the dark side of Lean: A systematic review of the lean factors that influence social outcomes. *The TQM Journal*, 33(2), 1754–2731.
Schmidt, A., Sousa-Zomer, T. T., Andrietta, J. M., & Cauchick-Miguel, P. A. (2018). Deploying Six Sigma practices to General Electric subsidiaries in a developing economy: An empirical analysis. *International Journal of Quality & Reliability Management*, 35(2), 446–462.
Sharma, S., & Chetiya, A. R. (2009). Simplifying the Six Sigma toolbox through application of Shainin DOE techniques. *Vikalpa*, 34(1), 13–29.
Singh, G., & Singh, D. (2020). Critical success factors for Six Sigma implementation: A systematic literature review. *Journal of Asia Business Studies*, 14(5), 795–818.
Six Sigma: A Complete Step-by-Step Guide: A Complete Training & Reference Guide for White Belts, Yellow Belts, Green Belts, and Black Belts. (2018). Council for Six Sigma Certification.
Snee, R. D. (2010). Lean Six Sigma – Getting better all the time. *International Journal of Lean Six Sigma*, 1(1), 9–29.
Taghizadegan, S. (2006). *Essentials of Lean Six Sigma*. Academic Press.
Tague, N. R. (2024). *The Quality Toolbox*. ASQ Quality Press.
Urban, W. (2019). TOC implementation in a medium-scale manufacturing system with diverse product routing. *Production & Manufacturing Research*, 7(1), 178–194.
Voehl, F., Harrington, H. J., Mignosa, C., & Charron, R. (2013). *The Lean Six Sigma Black Belt Handbook: Tools and Methods for Process Acceleration*. CRC Press.
Wickramasinghe, G. L. D., & Wickramasinghe, V. (2017). Implementation of lean production practices and manufacturing performance: The role of lean duration. *Journal of Manufacturing Technology Management*, 28(4), 531–550.
Womack, J. P., & Jones, D. T. (2003). *Lean Thinking: Banish Waste and Create Wealth in Your Corporation*. Simon & Schuster UK.

4

What Is the Lost Value: Muda and Constraints

4.1 General Concept

Everything in business starts with value. No matter what kind of development path the organization chooses or what kind of improvement methodology it employs, the questions of how value is produced and how value is destroyed are of primary importance. Value is the central concept for process improvement, and it always starts with the perception of the customer. Thus, the organization has to make appropriate efforts to understand value as its consumers and recipients understand it. This allows it to clearly point out how exactly value is produced in the organization and how value is lost.

The special kind of waste, called muda, explains comprehensively how the organization is losing value in the stream of organizational processes. There are eight kinds of muda; these are the eight main ways the organization wastes value. They are interdependent to some extent; however, each of them shows a separate mechanism by which value is leaking from the value stream. At the same time, muda is something that widely exists and is noticeable, but nobody dares to suppose it as waste. The concept of muda brings a refreshed view of process flow and allows an understanding of how much can be improved and streamlined in the processes and their coordination.

The reader learns about the basic concepts related to flow management in production systems. They learn the difference between a bottleneck and a constraint. The importance of correctly identifying constraints is emphasized. This is necessary in order to improve the throughput and efficiency of the entire company. Eliminating constraints leads to the main benefit of applying the Theory of Constraints — increased profit. The rest of this chapter presents selected methods for identifying bottlenecks in a production system. These are: process cycle time, process cycle time including production losses, and work-in-process inventory. The choice of bottleneck identification method must be adapted to the characteristics of the production system under study, but also to the data available to the manager. Each

DOI: 10.1201/9781003655237-4

of the methods described has its advantages and limitations, so there is no single solution for their selection. This chapter ends with a practical example describing the application of selected methods in identifying bottlenecks.

4.2 Introduction

Every organization strives to transform obtained resources into products or services that are delivered to customers and generate profit for the company as efficiently as possible. This is a circulation of value within the organization and between the organization and its partners. The organization tries to maximize the value perceived by customers while minimizing the value that is consumed to create value for customers. However, a huge portion of the value within the organization is wasted.

Improvement methodologies advocate the primary importance of value as the perception of customers obtaining the products delivered by the stream of organizational processes. Apart from this view on value, the value of human efforts and other resources consumed in the whole transformation stream—from materials and components to the fully functional product—is also important. The concept of muda implies that enormous amounts of resources and effort are wasted every day in organizational processes. If the organization can understand the extent of wasted resources and efforts, it can formulate bold improvement steps to achieve measurable and significant results.

In order to increase throughput in a production system, it is necessary to identify its bottleneck, that is, the factor that affects the increase in system capacity. As Eliyahu M. Goldratt noted, every organization has several potential bottlenecks, but at any given moment, only one bottleneck limits profit growth the most. Correct identification of this constraint is crucial because it allows improvement efforts to be focused where they will yield the greatest increase in throughput. Only conscious management of constraints improves the performance of the entire organization. Conscious management of constraints is one in which the role of the constraint in the system is known, the methods for identifying it are known, and the manager has the skills to analyze the data obtained, on the basis of which further decisions are made that will actually increase throughput.

4.3 Lost Value—The Concept of Muda

Value is a key notion in the Lean Management approach, but other methodologies also see value as something of primary importance. Value is a reference base for Lean principles. Value, as seen by customers, should be thoroughly known and systematically renewed. Additionally, the production stream should be simplified and streamlined to bring better value (Womack & Jones, 2003). For a proper understanding of waste in the organizational system, value must be considered.

In everyday use, value is primarily characterized by financial references. We talk about the value of a product, which we approximate as the price we have to pay for it. A higher price for a product means a higher market value, while a lower price means a lower market value. Financial value is linked to market exchange. In business, we talk about customer value, share value, or investment value. Value as money is the typical first association with the word value.

On the other side of this semantic continuum, so to speak, there is the very elusive meaning of value. These are values understood as a kind of ethical norm that is accepted in a given community and constitutes a point of reference regulating the behavior of the community. So, we are talking here about values in a much more abstract sense; they can also be transcendentals that determine the foundations of the functioning of societies.

However, the value understood as usefulness is the one that is most important in process improvement systems. In dictionaries, the concept of value is explained, among other things, as a characteristic of a thing according to which it is considered more or less desirable, useful, respectable, and important (YourDictionary, 2025). Value, understood as a degree of product usefulness in comparison with other products, is at the heart of marketing theory.

Marketing theory, in particular service marketing, draws attention to value. "Value in use" is mentioned there; according to value in use, the supplier does not create value but only a value proposition—an offer of value—and the value itself arises only when the product is used by the customer. During the use of the product, an experience arises that can be described as an experience of value (Grönroos & Ravald, 2011).

The concept of value in use implies the creation of value during product use, so the value depends not only on the product itself but also on the context of its use by the customer, such as the conditions in which it is used. Therefore, the full characteristics of the product, seen through the eyes of the consumer together with the context of its use, are the reference point for defining value in whole process improvement.

Illuminated by the understanding of value as above, an organization can qualify what constitutes value and what does not, and consequently what adds value and what does not. Typical for process improvement is the analysis toward distinguishing between activities within the organization that add value to the product and those that do not. The most popular practical criterion for determining whether any activity in the process stream adds value or not is the answer to a simple question: Would the customer be willing to pay for it? (Rother & Shook, 2009).

So obviously, the waste in the process stream cannot be understood and detected without exhaustive knowledge of what value is based on, and this applies to each specific product delivered by the organization to the market. If we allocate various resources to do something that does not produce value, we create waste, and this waste can also be estimated in lost money.

This lost value is called in the process improvement approach by the Japanese word muda. It is an important category of value that plays a significant role in the flow of organizational activities. This value is negative, that is, value that the organization loses. Therefore, it represents a specific perspective on value perception, which we usually interpret in a positive sense, as a benefit.

It can generally be assumed that everything in the process stream that does not contribute to the creation of value is a waste of resources for the organization. From this perspective, the main effort of process improvement should be focused on investigating waste and on methods to eliminate it afterward.

4.4 Types of Muda

The literature often refers to the list of seven types of muda presented by Taiichi Ohno, the main architect of the Toyota Production System. Ohno's insights are precious and still totally relevant. However, for their proper understanding it has to be underlined that muda has been observed in the mass production environment, automotive production specifically, which involves the large-scale manufacture of complex and technically advanced products. Furthermore, it was around the technological state of the 1980s and when economies were still reeling from the oil crises.

According to Ohno, becoming aware of muda and its identification is a preparatory step toward implementing the system of Toyota (Ohno, 1988, p. 42). Managers' awareness of what is muda along with the scale of wasted resources is indeed a logical starting point for changes in the process flow. Because if we do not realize how much value is lost in the company, we will probably never make truly important and bold changes.

Unfortunately, the author of the types of muda (Ohno, 1988) explains his types of waste only to a moderate extent, clearly considering them to be rather obvious, in the sense that they are unquestionably and obviously activities that are actually unnecessary, yet consume resources and lead to costs.

Wastes are related to work in the plant and the organization of process flows. If we assume that a certain job is necessary, for example, the production of a product with the minimum necessary effort, that is, doing only what creates value, then everything else is waste. In that case, waste can be interpreted as a reduction in the current production capacity of the plant.

Recognizing waste allows us to understand that it would be possible to produce a significantly larger number of products in a given unit of time with the given machines and employees of the plant. However, due to waste, significantly fewer products are produced. The types of waste identified by Ohno should be understood from this perspective.

Ohno lists overproduction waste as the first item on his list. "In business, there is no waste more terrible than overproduction" (Ohno, 1988, p. 28)—says the author of the concept of waste. The justification for the harmfulness of this type of waste is related to the typical mass production and large production batches in those days, with high inventory levels in all links of value chains.

The author justifies this with the prevailing way of thinking in business, similar to that of farmers: once the crops have been harvested, they must be stored because a disaster may occur and then these stocks will be used and possibly safe from the hunger. The author states that modern societies should have the courage, but also the common sense, to procure only what is needed, when it is needed, and in the quantities needed.

Producing exactly what is needed is characteristic of slimmed down flow of processes and this is exactly Just-in-Time logic, which the Toyota supply chains were famous of. The practice of insuring oneself with larger stocks at that time was a consequence of the oil crises. The muda of overproduction should be perceived as the stream of processes producing more than the customer is currently able and willing to receive.

Overproduction is a particular type of waste because it directly contradicts the idea of slimmed down value stream, which involves the continuous addition of value and immediate transfer of high-value products to the customer. Overproduction directly leads to other types of muda, especially waste of inventory.

Overproduction does not occur exclusively in connection with deliveries to external customers, where previously produced "overproduction" is stored by the company. Overproduction also occurs in individual processes that are coordinated in such a way as to produce in advance, rather than exactly what is currently needed for the next process in the stream.

Lean stream is one that is pull-based, that is, it operates in line with the needs of external and internal customers. The occurrence of overproduction muda proves the fundamental weakness of a company's operating system. Therefore, overproduction is the mother of all other muda, and because it has broad implications, it is not easy to measure its value in terms of costs.

From the abovementioned reasons, the muda of overproduction raises many questions and doubts among managers, especially in companies that are just starting their way of process improvement. One such question is whether it is really worthless building production inventories during periods of limited availability of components or materials, as this can bring significant benefits to the company. The answer may contradict the principles of eliminating the muda of overproduction.

Another concern is that if I give up producing for stock, I will not be able to meet the entire demand during the peak season, and the company will lose a lot. And yet another question is why a machine in a factory should only be used for part of the shift, as this is a real waste; let it produce "something" during that time. The concept of muda, along with muda of overproduction particularly, is in fact not about management strict rules but about what separates a company from lean operations flow in the long run.

It should be emphasized that an efficient and flexible flow, allowing for rapid scaling of production capacity to the scale of demand without generating additional costs, brings numerous benefits, the most important of which is the ability to perfectly respond to customer needs in terms of product quantity and functionality, which is usually of great value to them. Other benefits include the simultaneous elimination of many other types of waste that are closely, even functionally, related to overproduction.

Eliminating overproduction is the essence of streamlining the value stream, but it is also a huge challenge requiring managerial courage and a strategic vision for improving operational flow. Eliminating overproduction leads to low process variability, and low overproduction is only possible when processes are highly predictable.

In manufacturing, muda of overproduction is about eliminating excess physical products. However, overproduction occurs intensively in other types of operations and consists, for example, of generating unnecessary information or additional software functions in IT, creating reports or services that are not currently needed by the customer. This leads to sluggish service systems with low responsiveness to customers.

Waiting for work to be done is next type of muda. This mainly refers to operators waiting for a task to start or continue. Usually, when operators are waiting for certain reasons, other production resources, such as machines, are also waiting, meaning that they are not producing value at that time.

There can be many reasons for waiting. In relation to the previous muda, this could be, for example, poor harmonization of the flow of tasks and products in the stream of processes. However, the cause may be a lack of materials delivered to the workstation on time, a lack of information, such as the exact technical specifications. In addition, there may be a lack of tools to carry out the work, a lack of instructions from managers, machine breakdowns, or waiting for a product in progress from a previous workstation.

It should be emphasized that waiting is practically invisible to observers outside the workplace, as operators do not remain motionless in such situations at the plant or at their service spots. The actual scale of waiting can therefore come as a big surprise to managers. Waiting without a doubt has negative social effects within the organization, provoking everyone to pretend and fake work, almost automatically worsening identification with work and employee engagement.

The muda of waiting does not only occur in the production stream. Waste caused by waiting usually results from prolonged procedures and prolonged decisions blocking subsequent project steps. Delays caused by cross-team dependencies are quite often observed, as are those caused by waiting for approvals, confirmations, or reviews.

Prolonged waiting in one of the stages of the entire operation stream may be a symptom of a bottleneck occurring there. The time variable is one of the methods of determining the bottleneck of the system in the logic of the Theory of Constraints. But this issue will be discussed further in this chapter.

Transport is the next waste—a type of muda. Internal transport, moving materials and work in progress (WIP) through production halls and between facilities, does not add any value to the product being manufactured. However, it is costly, particularly in terms of the labor and the means of transport. In addition, parts are more likely to be damaged during transport, time is lost, and the entire flow is slowed down. Where much of transport exists, it implies the dispersed location of processes, which makes difficulties in flow of effective coordination.

The muda of transport also applies to the movement of tools, other items, or documentation. These activities consume resources and obviously do not add value; they can also contribute to other types of waste. Simple and short flows are the key to reducing waste of excessive transport, and they can be achieved by arranging workstations according to the sequence of process tasks.

In addition, every handoff is inherently associated with waste. Each transfer of work involves not only physical transport but also the risk of losing context and knowledge and what can lead to defects, which is another type of muda. Therefore, in service operation streams, eliminating excessive transport is not so much about optimizing the layout of workstations in rooms but about building cross-functional teams that reduce the need to transfer work.

Another damaging smooth flow type of muda is overprocessing, which commentators sometimes confuse with overproduction (discussed above). This type of muda involves processing a product to a degree that is not needed by the recipient; these are various functions of the product that the customer does not need but which are provided to them. The significance of this waste is enormous, because selected parts of the production process may be working completely unnecessarily.

Hence, the great importance of understanding value in terms of the customer's experience, as discussed above. Attention should be paid here to large projects where new solutions are created and where customers have no experience with using the future result of the project. Usually, there will be a lot of waste from overprocessing, that is, delivering a value proposition that will not translate into value for the customer. In IT services, doing more than the customer requires is referred to as "gold plating."

However, attention should also be paid to inappropriate processing methods in the process as another important manifestation of this waste. A common manifestation of such a waste is the introduction of various security features on the product during production, only to remove them in subsequent processes. This wastes the operators' work, prolongs time, and consumes the material used for security features.

Another source of overproduction waste is related to bureaucracy, lack of trust within the organization, and overly complicated procedures. This leads to too many approvals and acceptances, which are activities that do not add value and their presence in the processes significantly diminishes process productivity rates.

Inventory is another muda that, in the case of production, is easily noticeable when walking through production halls. Inventory is a potential value that has not yet reached the customer to become a real value. And yet, inventory has absorbed a certain value from the company, materials have been purchased, various production operations have already been carried out, and this value is not working further and does not generate revenue because the inventory is lying idle in the company.

Therefore, inventory stored in storage areas and warehouses constitutes frozen capital. The cost of frozen capital is determined by the relevant interest rates. Inventories typically occupy large areas within a plant, which costs appropriate amounts of money. Furthermore, inventory requires supervision and management. For example, plants often need to move inventory from a storage area at a workstation to deliver another WIP to this work area. This is, of course, an additional work, waste of time, and leads to complication of workflow.

Piling up inventory in warehouses often leads to damage to parts during storage, which results in additional losses. However, the most significant harm caused by inventory is that it hinders the coordination of flow, hiding shortages, delays, and poor synchronization of the value stream. Inventory makes it virtually impossible to identify other types of waste and thus eliminates them effectively.

In the sphere of service flows, inventory is also important and harmful. The service inventory manifests itself as queues, for example, with a stack of applications waiting to be processed. Waiting for unfinished projects is also a muda of inventories. Practitioners point out the inventory as the most important symptom of the level of value flow efficiency.

The Kanban system, for example, is designed to easily reduce work-in-process inventory levels, revealing the real problems in the flow—different kinds of waste. Hence, the great popularity of the Kanban method in manufacturing and other areas, especially in project management.

Unnecessary motion, that is, motion that does not create value for customers, also constitutes muda. This refers to motions performed by operators, such as unnecessary movement around the halls or around manufacturing machines. It also includes unnecessary manual movements performed by operators, such as carrying, moving, and reaching.

Movements that can lead to accidents or injury to operators are of particular importance, but others can lead to damage to equipment. Frequently encountered unnecessary motion occurs during machine changeovers, which is one of the reasons why Toyota developed SMED, a method to help streamline changeovers. All operator efforts should be focused on value-adding activities.

In services, unnecessary movement includes not only physical motion, such as walking, bending, or reaching, as mentioned above, but also unnecessary motion in the digital space and in terms of human cognitive attention. This refers to repeatedly switching between applications or digital environments, multitasking, and frequent context switching in human work.

Although this does not require physical effort, it forces the employee to mentally overload, which prolongs the completion of tasks, causes fatigue, and significantly increases the likelihood of errors. Thus, in services, it is not only about eliminating physical movement, but also about minimizing switching between tasks and between digital environments.

Defects as waste are listed by Ohno (1988) as the last on his list of seven types of muda. Defects in products and any errors in processes constitute wasted human and machine labor and wasted material, and require either repair or disposal as waste. Defects that reach customers are particularly negative, as they not only cause a loss of value for customers, but often a loss of customers themselves.

Defects are a very old topic in industry, and quality management offers a number of theories, systems, and methods for achieving zero-defect level. At the same time, however, the issue of process defects has not been resolved in modern business. What is more, there are still companies that operate on the assumption of an optimal level of defects, that is, a certain percentage of defects is accepted, for example, 0.2%, based on the assumption that achieving defect-free work, as older economic theories argued, would be cost-inefficient.

It should be clearly stated that every minor defect and employee mistake is a waste, so they should not occur at all. Defects can also sometimes be hidden at the workstations where they occur and then repaired at these workstations. But this is an even greater waste, because it distorts the actual production capacity of the workstations.

The value stream should strive to eliminate all defects. The absence of defects is a prerequisite for the efficient flow of the value stream, as defects lead to disruptions in the flow.

In service sectors, where the product is co-produced with the customer, defects may be intangible in nature and therefore more difficult to detect. However, the effects of mistakes can be just as severe and often even more destructive than in manufacturing. Due to the difficulty of standardizing services and the emerging nature of customer expectations, achieving zero defects in service products is a major challenge.

There is one additional muda that Ohno did not consider, namely, the muda of untapped employee talents. This refers to the failure to utilize the talents, knowledge, and experience of employees for the benefit of improving the value stream by allowing them to become disengaged in the organization.

The Gallup Institute is the world leader in monitoring employee engagement. Their research on large samples of employees shows that only 21% of employees worldwide are engaged in their work (Gallup, 2025, p. 6). The fundamental question concerns the value lost by companies as a result of the talent potential of 79% of employees, which is undoubtedly a huge waste.

The companies very often lose the great creative potential of its employees, who could perform their tasks much better. It should be emphasized that when work is based on repeating imposed routines and employees are deprived of the desire to best fulfill its purpose, it loses quality and productivity. Process improvement initiatives will be doomed to low effectiveness if employees implement them as a result of coercion.

Finally, employee engagement is an essential component of service product quality. Service products are co-created with customers, and service encounter staff personally create the product. Their attitudes and behaviors are perceived as valuable when they are genuinely engaged—both emotionally and mentally—in the creation process.

The process improvement perspective makes to see muda of untapped employee talents as a factor related to each of the other seven wastes. Additionally, Liker points out important causes of muda resulting from periodic work overload and work inequality as causes leading to wasteful work. Overload involves burdening people or machines with work to an excessive degree, and imbalances involve large disproportions in tasks between individual periods in schedules (Liker, 2004, p. 115).

The types of muda along with their key characteristics are listed below:

- Overproduction. It contradicts lean flow, enables and creates other types of waste.
- Waiting. When employees wait, no value is created, which also lowers team morale.
- Transport. Moving production around the plant causes many costs and losses, and it does not add value to the product.
- Overprocessing. Providing functionality or performing processes that have no value to customers.
- Inventory. Inventory costs money and hides problems; it is a measure of flow efficiency.
- Unnecessary motion. All operator efforts should create value; any other effort is wasteful.
- Defects. Every defect is wasteful and should be eliminated through prevention; zero defects are a prerequisite for good flow.
- Untapped employee talents. Employee creativity and commitment are the most important factors in a company's development; if systemically wasted, struggle to eliminate waste will be fruitless.

As shown above, many types of muda are interdependent, and types of waste often coexist. Muda, first formulated in the context of mass production, is fully applicable to every type of process flow. The concept of muda leads to a fundamental shift in thinking about the organization of operational flows. It makes us realize how much value is systematically wasted, which demonstrates how enormous the potential for improvement is.

The lens of muda also opens totally new opportunities when improving the system of processes in the flow at the bottleneck point. Regaining some productivity by eliminating muda leads to significant effects on the throughput of the entire flow. However, ensuring the stability of processes cooperating with the bottleneck undoubtedly involves eliminating muda from them. Considering what is demonstrated in previous chapters, it can be argued that muda and process variability are two sides of the same coin.

4.5 Why Correctly Identifying the Bottleneck Matters

The Theory of Constraints, as the name suggests, is based on constraints. More specifically, it involves identifying the limiting factor and then focusing on managing it in order to increase its throughput. The idea behind a manager's work is to find the constraint that currently limits the system's ability to achieve better results in relation to its goal (Mabin & Balderstone, 2003, p. 570). Initially, the Theory of Constraints was only applied in manufacturing, where the constraint was called a bottleneck. The clear and simple logic of the Theory of Constraints led other environments to try to implement the same thing, for example, in project management or distribution. At that point, the term bottleneck began to be misleading and had to be replaced with another, broader term, namely, constraint (Cox & Schleier, 2010, p. 149). However, in scientific research in the area of production management and in business language, these two terms are often used interchangeably. Therefore, it is worth discussing the authors' views on this subject.

There are many interpretations of bottlenecks in the available literature. Among the first researchers to analyze this issue were Lawrence and Buss (1994), who concluded that there is no clear consensus on the definition of a bottleneck. They based their findings on three definitions that were in use at the time:

1. A resource for which demand exceeds capacity.
2. A resource for which the waiting time for WIP is maximum.
3. A resource with maximum utilization.

Other researchers have also attempted to define bottlenecks. The literature contains terms such as:

4. the resource whose production rate is the lowest in relation to the entire production process (Kuo et al., 1996),
5. the resource whose average operating time is the longest (Roser et al., 2001),
6. any resource whose production capacity is equal to or less than the demand for it (Goldratt & Cox, 2004),
7. the resource with the lowest productivity, thus limiting the throughput of the system (Betterton & Cox, 2009),
8. a critical resource of the system that determines its performance (Szatkowski, 2014),
9. the resource that does not have sufficient capacity to cope with actual demand (Ashlag, 2015),
10. the weakest link, limiting the full utilization of the capabilities of a company or production system; it prevents development by inhibiting subsequent activities in the supply chain (Pająk, 2021),
11. the process affecting throughput and delivery time (Thürer et al., 2021).

Discussing the above considerations of researchers, a bottleneck is an element that limits system performance or an element associated with a specific metric that is related to the selected method or methods of identifying bottlenecks.

Furthermore, the concept of limitation can also be found in scientific research. It can be explained as:

1. anything that prevents the system from achieving greater effectiveness in achieving its goal (Corbett, 2009),
2. a cause that has adverse effects on the system (Trojanowska, 2011),
3. the problem located in any area of the enterprise (Cyplik et al., 2012),
4. obstacles that stand in the way of achieving greater profit (Sims & Wan, 2017),
5. an element of the system that limits its operation and often hinders the achievement of its intended goal (Lewandowski et al., 2018),
6. anything that prevents the system from achieving better results in relation to its goal (Scheinkopf, 2022).

According to researchers, the limitation is quite general and very broad. It refers to any problem that prevents a company from improving its performance.

Looking at how bottleneck and constraint are defined shows that it's worth taking a closer look at what each really means and how this understanding affects their practical use. The bottleneck is an internal factor of the production system that limits the growth of production flow. At any given moment, this factor can be found anywhere in the entire production stream—from the moment the raw materials are received to the moment the finished product is delivered to the customer. A constraint is a factor limiting the full utilization of a company's capabilities, which occurs in various areas of the company as well as in its environment. A constraint is a factor that limits the company's development at a given moment. Accepting these interpretations of bottleneck and constraint allows us to draw another important conclusion. In other words, a bottleneck can be seen as a special type of constraint.

Now that we understand what a constraint is, we can move on to the next stage—its identification. As explained earlier, the Theory of Constraints, in its initial stage, is based on identifying the limiting factor. A resource that restricts production capacity is referred to in the literature as a Capacity Constraint Resource (CCR). The CCR can be a machine, a process, or an area with the lowest processing capacity. E. M. Goldratt argues that every company has at least one constraint. If nothing constrained a company, it would achieve infinite value. The basic assumption of the Theory of Constraints is that the throughput capacity of an enterprise's entire production system is determined by the throughput capacity of the constraint (Goldratt, 1990, p. 4). It is important to understand what is preventing the system from achieving its goal, which is to make money today and in the future. It is important to find the so-called global constraint, that is, the one that has the greatest impact on the company's goal and will bring the greatest results. Importantly, improving the so-called local efficiency, that is, minor improvements in the company, will not affect the throughput of the entire system.

Correctly identifying bottlenecks is key to improving a company's throughput. Throughput is the money a company earns from sales, minus its fully variable costs. Improving throughput is considered the most important way to increase a company's profitability (Urban & Rogowska, 2018).

The Theory of Constraints assumes that the production capacity of an entire enterprise system depends on the processing capacity of its weakest link—the constraint. This is a factor that determines the production capabilities of an enterprise, but at the same time, it is a factor that can limit its success, which is why proper management of it is extremely important. Identifying the constraint is key to improving the production system and increasing profits. Companies that are constantly developing and striving for continuous improvement strengthen their competitive position in the market. E. M. Goldratt argues that there are undisclosed reserves of potential in companies which, if properly exploited, will allow for significant growth, often without investment. Therefore, correct identification of constraints is necessary for proper company management (Dettmer, 2001).

According to E. M. Goldratt, companies that are constantly improving should effectively identify constraints. The occurrence of bottlenecks contributes to many negative effects that impact the organization. These include, for example, work-in-process inventory, downtime, order fulfillment delays, or problems with meeting delivery deadlines, and many others (Schultheiss & Kreutzfeldt, 2009). The available literature offers many methods for identifying bottlenecks. These methods are mainly based on observing the production system, analyzing quantitative data, or measuring machine operating time.

One of the basic methods of identifying bottlenecks is to analyze process cycle times (C/T process). This involves measuring the time it takes to produce a single unit of a product, for example, at a single workstation, machine, or process. Cycle time can be expressed in seconds, minutes, hours, or even days per unit. Identifying bottlenecks involves comparing all the cycle times collected and determining which one takes the longest and thus restricts flow. The stage with the longest cycle time indicates the bottleneck in the production system under study. This is where resources should be concentrated in order to eliminate the bottleneck. This approach indicates the maximum efficiency of processes under the conditions studied (Roser et al., 2015, p. 64). In practice, cycle time measurement depends on several factors related to the nature of the process and the organization of production at the facility under study. In some factories, it can be measured by standing directly next to the machine. Some companies have automatic machine time recorders and production stage tracking systems, so cycle time can be determined based on the collected data.

The C/T process method is supplemented or broadened by an analysis of process cycle time that takes into account the so-called production losses. The analysis should consider not only the nominal production time per unit, but also other factors that cause a loss of efficiency. What are production losses? These are all factors that directly prolong the processing of a batch. These factors may result, for example, from waiting for the availability of machines or processes, breakdowns, machine changeovers, or unplanned downtime, which prevent or limit the ability to produce products at a given time. This method is ultimately a good approach because it takes into account the real conditions of production. In every plant, there are reasons why it is not possible to manufacture products at a given time. It can therefore be considered that this method (compared to the basic process cycle time method) is more suited to the actual circumstances of the manufacturing process. Loss time should be determined using data recorded by the company. Taking production loss time into account in the C/T value of processes is an important element that ultimately affects the result. It gives a more realistic picture of the throughput of a given process. Moreover, by identifying production losses, this method allows for better selection of improvement measures in the next stage. According to this method, the process with the longest C/T is the bottleneck (Urban & Rogowska, 2020, p. 75).

Another commonly used method is the analysis of WIP inventory. This involves visually assessing the inventory status before the process and performing a detailed analysis of its levels. The flow of production in the manufacturing system is linked to WIP awaiting processing in subsequent stages. It is necessary to conduct observations during which one should move along the process line. A high level of WIP is a sign that there is a bottleneck behind these inventories (Urban, 2019, p. 183; Roser et al., 2015). Importantly, in order to be able to identify bottleneck, it is recommended to make several observations (at least two) at different times, under typical working conditions in production. This will ensure that the collected WIP data is reliable.

A number of other methods for identifying bottlenecks can be found in the literature. In a study conducted by Urban (2019, p.184), the author developed a method of C/T corrected by process engagement. Another well-known and fundamental method is to compare takt time and resource cycle time (Sims & Wan, 2017) and others.

The method chosen should be individually tailored to the characteristics of the production system being analyzed and the information available. In addition, it is worth preceding the search for bottlenecks with a thorough understanding of the production flow. It is also worth learning about all the characteristics associated with bottlenecks. For example, understanding their efficiency allows you to understand what the problem is. By understanding the bottleneck and how it differs from other processes in terms of performance, a company can choose the appropriate way to increase throughput through the bottleneck. You can read more about detecting bottlenecks in Chapter 6 of this book.

4.6 The Bottleneck Identification—A Business Case Study

The research object is a medium-sized manufacturing company in the printing industry. The company produces printed packaging for the food and pharmaceutical industries. The company's production system is typical serial production (it produces small and medium batches). It is characterized by varying process loads, which means that not all production processes are always involved in the manufacture of a product. The production process consists of four steps: (1) offset printing, (2) die stamping foil using the hot-stamping method, (3) die cutting, and (4) folding with gluing. Two methods were used to identify bottlenecks. The first was the C/T process. Direct measurement of cycle times at the machine was not possible due to the very short cycle times of each process, so the average cycle times for each process were calculated based on the number of printing sheets and their production time in a single month. The second method was WIP inventory. The inventory was calculated before each process. The data was collected by directly

TABLE 4.1

Measured characteristics of the production system

Process	Cycle time [sec]	Inventory waits for a process [pallets]
offset printing	0.31	57
die stamping	1.15	30
die cutting	0.71	15
folding with gluing	0.80	10

counting the pallets waiting in the production hall. The calculations were performed over two work shifts.

The findings presented in the table indicate that, according to the C/T process method, the bottleneck is the die stamping process. However, the WIP inventory method indicates that the bottleneck is the offset printing process. The next step is to take a closer look at the results obtained in order to correctly identify the bottleneck. It is worth mentioning here that only a correctly identified bottleneck, that is, one that has the greatest impact on the company's goal, will bring the greatest results. It has been established that printing sheets (cardboard) are purchased and delivered from external manufacturers on specific terms. For safety reasons, the company under investigation keeps high stocks of raw materials. Therefore, it should not be indicated that the bottleneck is the offset printing process. It has been established that the bottleneck is the die stamping process. The next step indicated by Goldratt and Cox (1992) is to look for opportunities to exploit the bottleneck.

References

Ashlag, Y. (2015). *Zasady TOC: Zarządzanie ograniczeniami dla wzrostu biznesu.* MintBooks.

Betterton, C. E., & Cox, J. F. (2009). Espoused drum-buffer-rope flow control in serial lines: A comparative study of simulation models. *International Journal of Production Economics*, 117(1), 66–79.

Corbett, T. (2009). *Finanse do góry nogami: zdroworozsądkowa rewolucja w rachunkowości* (2nd ed.). MintBooks.

Cox, J. F., & Schleier, J. G. (2010). *Theory of Constraints Handbook.* McGraw-Hill Education Ltd.

Cyplik, P., Adamczak, M., & Hadaś, L. (2012). Critical chain project management and drum-buffer-rope tools integration in construction industry – case study. *Log-Forum*, 8(1), 29–37.

Dettmer, W. H. (2001). *Beyond Lean Manufacturing: Combining Lean and the Theory of Constraints for Higher Performance.* System International.

Gallup. (2025). *State of the Global Workplace*. Understanding Employees, Informing Leaders. 2025 Report.
Goldratt, E. M. (1990). *What Is This Thing Called the Theory of Constraints and How Should It Be Implemented?* North River Press.
Goldratt, E. M., & Cox, J. F. (1992). *The Goal: A Process of Ongoing Improvement* (2nd ed.). North River Press.
Goldratt, E. M., & Cox, J. F. (2004). *The Goal: A Process of Ongoing Improvement* (3rd ed.). North River Press.
Grönroos, C., & Ravald, A. (2011). Service as business logic: Implications for value creation and marketing. *Journal of Service Management*, 22(1), 5–22.
Kuo, C. T., Lim, J. T., & Meerkov, S. M. (1996). Bottlenecks in serial production lines: A system-theoretic approach. *Mathematical Problems in Engineering*, 2, 233–276.
Lawrence, S. R., & Buss, A. H. (1994). Shifting production bottlenecks: Causes, cures, and conundrums. *Production and Operations Management*, 3(1), 21–37.
Lewandowski, J., Skołud, B., & Plinta, D. (2018). *Organizacja systemów produkcyjnych*. Polskie Wydawnictwo Ekonomiczne.
Liker, J. K. (2004). *The Toyota Way: 14 Management Principles from the World's Greatest Manufacturer*. McGraw-Hill.
Mabin, V. J., & Balderstone, S. J. (2003). The performance of the theory of constraints methodology – Analysis and discussion of successful TOC applications. *International Journal of Operations & Production Management*, 23(6), 568–595.
Ohno, T. (1988). *Toyota Production System: Beyond Large Scale Production*. Productivity Press.
Pająk, E. (2021). *Zarządzanie produkcją*. Wydawnictwo Naukowe PWN.
Roser, C., Lorentzen, K., & Deuse, J. (2015). Reliable shop floor bottleneck detection for flow lines through process and inventory observations: The bottleneck walk. *Logistic Research*, 8(7), 1–9.
Roser, C., Nakano, M., & Tanaka, M. (2001). A practical bottleneck detection method. *NJ: Institute of Electrical and Electronics Engineers*, 2, 949–953.
Rother, M., & Shook, J. (2009). *Learning to See: VSM to Create Value and Eliminate Muda*. Lean Enterprise Institute.
Scheinkopf, L. (2022). *Narzędzia myślowe TOC dla managerów: Logika zmiany*. MintBooks.
Schultheiss, J., & Kreutzfeldt, J. (2009). Performance improvement in production systems through practice-oriented bottleneck management. *Proceedings of the 4th European Conference*, 2(1), 23–38.
Sims, T., & Wan, H. (2017). Constraint identification techniques for lean manufacturing systems. *Robotics and Computer-Integrated Manufacturing*, 43, 50–58.
Szatkowski, K. (2014). *Nowoczesne zarządzanie produkcją*. Wydawnictwo Naukowe PWN.
Thürer, M., Ma, L., Stevenson, M., & Roser, C. (2021). Bottleneck detection in high-variety make-to-order shops with complex routings: An assessment by simulation. *Production Planning and Control*, 33(15), 1–12.
Trojanowska, J. (2011). *Zarządzanie produkcją na zamówienie w oparciu o teorię ograniczeń. Wybrane problemy logistyki produkcji*. Wydawnictwo Politechniki Poznańskiej.
Urban, W. (2019). TOC implementation in a medium-scale manufacturing system with diverse product routing. *Production & Manufacturing Research*, 7(1), 178–194.
Urban, W., & Rogowska, P. (2018). The case study of bottlenecks identification for practical implementation to the Theory of Constraints. In *XV International Conference: Multidisciplinary Aspects of Production Engineering (MAPE), Conference Proceedings*, 1(1). Wydawnictwo PANOVA.

Urban, W., & Rogowska, P. (2020). Methodology for bottleneck identification in a production system when implementing TOC. *Engineering Management in Production and Services*, 12(2), 74–82.

Womack, J. P., & Jones, D. T. (2003). *Lean Thinking: Banish Waste and Create Wealth in Your Corporation*. Simon & Schuster UK.

YourDictionary. (2025, September). Retrieved from https://www.yourdictionary.com/

5

Constraints from Selected Industry Perspective

5.1 General Concept

Every company faces at least one constraint at any given time that affects its daily operations, growth, and increased throughput. Constraints in businesses prevent them from reaching their full potential. Identifying and managing constraints can bring many benefits to any organization.

This chapter presents an analysis of the constraints faced by the printing industry. The results of a survey conducted on a representative sample of 237 companies in the country are presented. The choice of the printing industry was deliberate. Printing companies are distinguished by a wide range of products, which makes identifying constraints a difficult task, but also an interesting area for research on detecting and eliminating these constraints.

The study identified a total of 26 different constraints in the printing industry. In each size category of enterprises—small, medium, and large—the three most frequently mentioned constraints were selected. The rest of this chapter describes the classification of the 26 constraints identified, which were divided into five groups: operational, market, organizational, managerial, and social. Each group contains between three and eight constraints.

The classification of constraints is intended to help printing companies identify the problems they may face and understand the significance of each of them. In addition, it enables companies to more precisely determine which constraints require special attention and to develop improvement measures aimed at partially reducing or completely eliminating them. This chapter concludes with an analysis of the correlation between company size and the occurrence of potential constraints.

DOI: 10.1201/9781003655237-5

5.2 Introduction

In today's world, companies operate in a highly dynamic and complex environment, and effective flow management is a key factor in a company's success. Regardless of the industry in which companies operate, each of them encounters various constraints that affect their daily operations, production potential, and strategy. The constraint itself encompasses a fairly broad perspective. It can be any element or area within or outside the company. Therefore, understanding the constraints faced by a given industry provides a general overview of what constraints may affect throughput growth and what constraints to look for in your own companies.

The printing industry is an interesting subject for research in terms of identifying constraints. Its production system is organized as a typical batch production (with low repeatability of batches), there is no continuous or even partial flow, and machine changeovers are time-consuming. However, it is worth noting that these characteristics are not unique to the printing industry. These are characteristics of multiproduct manufacturing organizations that are found in many other industries, such as furniture and metalworking. Identifying constraints for the printing industry can also be useful for other sectors with similar production organizations.

5.3 Constraints in the Printing Industry—Research Results

Identifying constraints and focusing on eliminating them is the guiding principle of the theory of constraints. According to Goldratt (1984), every company has at least one constraint that hinders the achievement of its intended goal at a given moment (more in Section 4.5).

Constraints occurring in companies hinder or even inhibit their development to varying degrees, preventing them from fully exploiting their potential. Constraint management allows for the concentration of improvement activities in areas that actually require it and have the greatest impact on the company's performance. Proper identification and understanding of constraints is crucial when making managerial decisions about corrective actions where they can actually have the greatest effect on the entire production system.

The object of the research is the printing industry. It is one of the fastest growing industries in the country in recent years. Economic changes affecting the competitiveness of companies in the printing industry are forcing manufacturers to seek solutions to streamline production processes. The production system in the printing industry is characterized by a high diversity of production orders, low repeatability of production series, and short production orders. Among other things, this results in frequent machine changeovers. Setting the correct print color and other machine parameters requires a long time and many attempts. The production system in the printing industry is also characterized by a lack of continuity of flow. This means that not all production processes are always involved in the production of the product range. With production organized in this way, identifying constraints is particularly important for ensuring smooth production and maintaining a stable market position. That is why it is worth knowing what constraints the printing industry faces. From the point of view of all printing companies, knowledge about constraints is valuable. It will indicate areas where constraints arise, enabling companies to identify similar problems in their own operations.

The survey was conducted among small, medium, and large printing companies. The number of entities surveyed was 237, which met the requirements for a statistically representative sample size for the country. The survey was conducted using the CAWI methodology supported by the CATI methodology. The respondents in the companies were senior and middle managers, and the survey was conducted over a period of three months.

The survey referred to constraints that occur in the printing industry. The question listed 16 different constraints, and respondents were asked to rate each of them using a five-point Likert scale. The question also included the option to indicate their own constraints that were not included in the question but may be relevant to the issue under analysis. The entire study covered a broader range of topics, but this chapter presents only a selected fragment relating to the analysis of limitations in the printing industry. The output of the study is shown in Table 5.1.

The results of the survey presented in Table 5.1 show that all the constraints identified in the questionnaire were assessed by respondents as occurring in the printing industry. This means that none of the areas included in the survey has been eliminated, and each of them may constitute a potential constraint affecting the development of the printing industry. The most frequently indicated constraints are the scale of demand for products (57%), applicable legal standards (54%), and machine capacity (51%).

Ten entities participating in the survey provided their own answers. Respondents listed the following factors as possible limitations for the company:

- quality of courier deliveries,
- insufficient storage space (production materials and finished products),
- customer requirements,
- availability of courier services,
- large number of changes between calculation and the final product,
- incompetence of employees,
- employee commitment to their work,

TABLE 5.1

Constraints in the printing industry

Constraint	Indications [%]
Scale of demand for products	57
Applicable legal standards	54
Limited number (monopoly) of suppliers	41
Conflicts between different departments	18
Incorrect flow of information	22
Lack of synchronization between the sales and production departments	21
Insufficient production space	29
Company development policy	25
Excessive product range diversity	20
Machine capacity	51
Outdated technology	24
Inappropriate technology	15
Production planning system	31
Delays in the delivery of materials to the workstation	33
Problems with the delivery of raw materials from suppliers	45
Unavailability of employees	32

- delays in files from customers,
- company owner, and
- high capital intensity of the company.

The constraints mentioned by respondents cover various areas. Some of the responses refer to typical external constraints, such as the quality of courier deliveries or customer requirements, or social constraints, such as employee incompetence and commitment. Interestingly, some respondents pointed to management-related constraints, such as decisions made by the business owner. This may indicate a growing awareness among managers of the impact of strategic decisions on the effectiveness of a company's operations. The survey results indicate a certain multidimensionality of the constraints faced by printing companies. It can be assumed that the constraints in the printing industry are complex and systemic in nature. They may involve internal areas of the company's operations or may occur in its external environment.

Analysis of the survey results revealed several limitations most frequently indicated by companies of various sizes—small, medium, and large. Table 5.2 presents the results of the analysis.

The results of the survey presented in Table 5.2 show that regardless of their size, printing companies indicate that insufficient demand for their

TABLE 5.2

The size of the company and the type of potential constraints

The size of the company	Constraint	Indications [%]
Small	Applicable legal standards	60
	Scale of demand for products	58
	Machine capacity	48
Medium	Scale of demand for products	57
	Applicable legal standards	54
	Problems with the delivery of raw materials from suppliers	53
Large	Machine capacity	80
	Applicable legal standards	60
	Scale of demand for products	60
	Limited number (monopoly) of suppliers	60

products is one of the key constraints hindering their development. This constraint may result from fairly high competition on the market, but also from the seasonality of orders. There are at least two periods of seasonality in the printing industry, that is, periods of increased demand for products. These are the periods of preparation for the holiday season and the spring-summer period. During these periods, the load on production resources is quite high, while in the remaining months of the year it remains at a much lower level.

Another constraint, regardless of the size of the company, is the applicable legal standards. The printing industry is subject to numerous national and EU regulations, including environmental protection requirements and the associated costs. Printing companies must comply with strict standards regarding harmful emissions, waste segregation, etc. Another requirement concerns copyright protection. The printing industry often uses images, photographs, texts, and other elements that are protected by copyright. The applicable legal standards may therefore constitute a constraint for companies due to the need to comply with their provisions, which may require additional time and financial resources.

Machine capacity is the constraint most often cited by large companies, but it is also one of the main constraints for small businesses. Large companies often specialize in printing specific types of products, which may be related to higher customer requirements and high product quality. This can result in a greater number of complex orders requiring diverse and more advanced processes, as well as the coordination of multiple machines and employees. This means that there are more machine changeovers, which in turn increases downtime and affects their overall efficiency. Although large companies have greater financial resources that allow them to invest in modern machinery, these factors can affect the efficiency of production processes.

Small businesses manufacture different types of products, serve customers with other requirements, and operate on a smaller scale, which is related to a smaller number of orders. They usually carry out less complex production processes and have less advanced machinery, which can affect capacity.

Medium-sized companies indicated constraints related to problems with raw material supplies from suppliers. The survey results showed that small and large companies also identify such constraints, but not to the same extent as medium-sized companies. It is possible that the differences in the findings result from a more complex supply structure and greater demand for raw materials than small enterprises, but at the same time, medium-sized entities do not have the same purchasing power as large companies. Large companies operate on a larger scale. This allows them to negotiate more favorable prices and better terms for the supply of raw materials. In addition, they may be able to secure adequate stock levels, which gives them an advantage over small and medium-sized entities. Small enterprises, on the other hand, usually do not have such a high demand for raw materials and can use smaller, local suppliers, which reduces the risk of supply problems.

The large companies have indicated that one of the main constraints is the monopoly of suppliers. Large printing companies have a greater demand for raw materials and supplies than small and medium-sized companies, so they may find it difficult to find alternative suppliers who can meet their needs. This constraint may also apply to the monopoly of suppliers of specialized machinery and technology.

5.4 Classification of Constraints in the Printing Industry

Research conducted on a representative sample of printing companies identified 26 different constraints that affect their daily operations and development. To better understand their significance, they were divided into five groups: operational, market, organizational, managerial, and social. Each group contains between three and eight constraints. The constraints were ranked according to the frequency of respondents' indications. The classification of constraints is intended to help printing companies understand the problems they may face and the significance of each of them. In addition, companies can more precisely determine which of them require attention and the development of improvement measures aimed at their partial reduction or complete elimination. The output of the study is shown in Table 5.3.

The first group consists of operational constraints. These result from the physical capabilities of the company and the availability of technology. They hinder the proper functioning of production processes, resulting from imperfections in the technologies and equipment used, as well as insufficient

TABLE 5.3

Classification of constraints in the printing industry

Group	Constraint
Operational	Machine capacity
	Insufficient production space
	Delays in the delivery of materials to the workstation
	Excessive product range diversity
	Outdated technology
	Inappropriate technology
	Quality of outsourced logistics services
	Insufficient storage space
Market	Scale of demand for products
	Applicable legal standards
	Monopoly of suppliers
	Customer requirements
	Availability of logistics services
	Problems with the delivery of raw materials from suppliers
	Large number of changes between calculation and the final product
Organizational	Production planning system
	Lack of synchronization between the sales and production departments
	Incorrect flow of information
	Delays in receiving files from customers
Managerial	Company development policy
	Business owner's management style
	High capital intensity of the company
Social	Unavailability/shortage of employees
	Internal conflicts between company departments
	Incompetence of employees
	Low employee engagement

space and infrastructure. They may relate to the production process, management systems, or transport. Operational constraints have a direct impact on, among other things, production capacity, product quality, and the speed of production order fulfillment. There are eight constraints in the group of operational constraints.

Machine capacity is one of the most frequently cited limitations in the printing industry, often referred to as a bottleneck. It means that a machine is capable of producing less than other machines. The constraint associated with low capacity affects many areas of a company's operations. Examples include the accumulation of work in progress, delays in the completion of production orders, and higher unit costs. Ignoring such limitations can lead to reduced competitiveness in the market (in cases where the customer wants a quick turnaround) and higher operating costs for inefficient machines.

Insufficient production space means that the company has too little space to carry out manufacturing processes. In the event of an increased number of orders, this limitation hinders the free flow of semifinished and finished products, leading to organizational conflicts between the various stages of production. Expanding the company's product range can also be a challenge due to the lack of space for installing new machines.

Delays in the delivery of materials to the workstation mean difficulties with the regular and timely supply of raw materials needed to fulfill orders. As a result, unnecessary downtime occurs and employees are unable to complete orders on time. The cause of the limitation may be external disruptions in the supply chain, but also internal organizational errors.

A large/excessive product range diversity means that a company offers a wide range of products that differ in shape, color, etc. Production variability results in frequent and usually time-consuming machine changeovers and the adjustment of appropriate machine parameters to the production process. This reduces production capacity, as the machines cannot manufacture products during this time. Frequent changeovers lead to downtime, longer order fulfillment times, and higher unit costs.

Outdated or inappropriate technology means that a company uses technology that is not modern and does not provide the same quality or efficiency as newer solutions, or that the technology is not suitable for the needs of the production process and the specific nature of the orders being carried out. This may include machines with low efficiency, high failure rates, or limited technological capabilities. Such limitations may, among other things, make it difficult to achieve the expected results, which in extreme cases may result in non-compliance with customer requirements. They may also increase production time, which leads to delays in the execution of production orders. In addition, mismatched technology reduces the competitiveness of the company. The company is limited in its ability to implement new products and has limited capabilities to handle complex or individual orders.

The printing industry largely outsources some of its logistics activities to external companies. The constraint on the quality of outsourced logistics services means that the company has problems with the quality of courier/transport services. These may include delays in delivery, damage to shipments in transit, or loss of this shipments. The use of subcontractors carries certain risks. It can limit control over the timeliness of order fulfillment, quality, and delivery security. Ultimately, this constraint can also lead to the loss of customers or a decline in their trust.

The last constraint in this group is the lack of sufficient storage space for materials needed for production and finished products (insufficient storage space). In other words, the company has limited space to store raw materials needed for production and finished products. This constraint can greatly hinder production organization and also affect the quality of stored raw materials by failing to provide adequate storage conditions. The consequence of this constraint is the need to frequently move raw materials and the risk of damage. In addition, the constraint may make it difficult to ensure a smooth flow and, in the long term, may cause delays in the delivery of the finished product to customers.

The second group consists of market constraints. These result from market conditions (external factors). They hinder the proper functioning of a company due to differences in product demand, legal standards, availability of logistics services, customer requirements, and the number of suppliers on the market. They may affect various aspects of a company's operations, such as production, distribution, or sales. Market constraints shape the conditions under which a company operates and often force it to adapt to new conditions. There are seven constraints in the group of market constraints.

The scale of demand for products is the most frequently cited constraint among market constraints. It means that the number of orders from customers is too small in relation to the company's capacity. In such cases, companies do not fully utilize their production potential. This constraint leads, among other things, to low capacity utilization and reduces the company's profitability. In the long term, this constraint also affects investment decisions and plans for the further development of the company. Low demand weakens the company's position in the market, which has a significant impact on its competitiveness.

Another constraint is the applicable legal standards. Printing companies are subject to numerous legal regulations and provisions concerning various aspects of their operations. These may include legal provisions related to environmental protection, waste management, pollutant emissions, or copyright. Another important area is compliance with requirements regarding the paper used in the printing process. The paper must meet technical standards, that is, weight, smoothness, whiteness, and others. Restrictions related to legal standards can pose financial and organizational challenges, especially for small businesses. For administrative reasons, they can also lead to longer production times.

Another constraint is supplier monopoly. This means that the company is dependent on a small number of manufacturers or distributors who supply paper, printing inks, offset plates, and other consumables. The monopoly of suppliers limits the ability to negotiate terms of cooperation and prices, which ultimately affects the final unit price of the product.

Customer requirements are a constraint faced by the printing industry. This means that customer specifications, flexibility in terms of personalization or format, maintaining the required quality, and meeting deadlines can become a challenge for printing companies. Today's customers expect high-quality service, the ability to make frequent changes to designs, and short turnaround times. Customer requirements can have a significant impact on production and organizational processes. Customer requirements are associated with the need to invest in modern technologies, improve employee qualifications, and adapt logistics services. This can affect the unit cost of a product and production errors resulting from time pressure. On the other hand, failure to meet customer requirements can lead to customer loss and a decrease in the company's attractiveness.

The availability of logistics services is another market constraint. The printing industry serves customers from different parts of the country as well as customers abroad. Therefore, it is quite heavily dependent on efficient external logistics services. This constraint results from an insufficient number of carriers who could deliver products to customers at short notice, high transport costs, and territorial restrictions. The limited availability of logistics services may affect the timeliness of product delivery to the customer, as well as customer loss or dissatisfaction.

Problems with raw material deliveries from suppliers mean that the raw materials needed for production are not delivered on time. This can have serious consequences for printing companies. If a company does not have sufficient stocks of the required raw materials and consumables, this may result in production downtime, which in turn will affect the timely completion of orders.

A large number of changes between the calculation and the final product means that during the product design process, the customer often modifies its parameters, for example, in terms of color. The production process cannot start until the customer finally approves the design. Frequent changes disrupt production planning and also necessitate recalculating costs. This constraint extends the order fulfillment time. In the long term, this reduces the organizational efficiency of the company.

The third group consists of organizational constraints. These result from errors in the internal functioning of the company and the organizational solutions adopted. They hinder the timely completion of production orders, the ability to respond quickly to changes, and internal communication. They can affect any area of the company. There are four constraints in the organizational constraints group.

The production planning system means that the current method of production planning in the company does not ensure optimal use of resources, coordination of production processes, or appropriate sequencing of production orders to avoid time-consuming machine changeovers. A production planning system that operates incorrectly can cause problems with product shortages or surpluses, work in progress, and inefficient use of production capacity.

A lack of synchronization between the production and sales departments means a lack of consistency between these departments. The information provided is incomplete and delayed, which ultimately affects the proper planning and fulfillment of orders. This can lead to untimely order fulfillment or make production impossible with the available resources. The lack of cooperation between these two departments may also limit the ability to respond flexibly to customer needs and changes.

Incorrect information flow refers to the lack of proper information flow between different departments of the company. Information flows in an untimely and disorderly manner, which causes communication errors. This constraint can lead to errors in production or incorrect or delayed decisions, as well as tensions between employees, which negatively affects the atmosphere in the organization.

Delays in receiving files from customers mean that customers do not deliver files, documentation, or other materials necessary to start the order/production on time. Among other things, this affects delays in relation to previously agreed deadlines. It can also lead to machine downtime, disrupting the production schedule. This forces changes to the order fulfillment plan.

The fourth group consists of managerial constraints. These result from the way the company is managed and the strategic decisions that are made. They hinder the full potential of the company and its resources from being realized. Managerial constraints may relate to development policy, poor investment decisions, or inappropriate management style on the part of the owner. Managerial constraints affect the long-term development of the company. There are three constraints in the managerial constraints group.

A company's development policy refers to the way in which senior management determines the company's growth potential and the utilization of its full potential. If this limitation affects the company, it means that it is probably not fully exploiting its potential, is not investing properly in modern technologies, and is not improving the qualifications of its employees. Inappropriate strategic priorities or a lack thereof can lead to difficulties in achieving business goals and a loss of competitive advantage. In addition, the company is not flexible in responding to customer needs and adapting to new technological trends. Eliminating this constraint should primarily involve defining the company's strategic priorities and investments consistent with its financial capabilities and, importantly, market expectations.

The owner's management style means that the most important decisions of the company, such as strategy, investments, or operational activities, depend on one or several managers. The constraint imposed by the company owner and his management style is the expectation of quick and high profits while at the same time unwillingness to invest financial resources. The development of a company involves frequent risk-taking, the owner's experience and knowledge based on market analysis, rather than making decisions based on intuition. This constraint makes it impossible to introduce changes. It hinders the implementation of new methods of managing the organization and, mainly, production, and also slows down the overall development of the company. According to Ashlag (2015, p. 46), the constraint associated with the style of business management is a constraint that is very difficult to eliminate.

The high capital intensity of a company relates to the costs associated with running a business. This means, among other things, that production requires significant financial outlays, especially in terms of purchasing and maintaining specialized machinery, consumables, and skilled workers. The payback period for such investments is often long, so companies cannot always afford to respond quickly to market changes or implement new technologies. If a company does not obtain external sources of financing, it usually postpones investing in new solutions and limits itself to modernizing its existing machinery. This constraint may affect the speed of order fulfillment, the attractiveness of the offer, or competitiveness on the market.

The last group consists of social constraints. These mainly result from human factors and employee attitudes. They hinder the performance of duties and lead to irregularities in the functioning of the company. They may concern interpersonal relations, employee commitment, their availability, appropriate qualifications, and cooperation between teams. There are four types of constraints in this group.

Unavailability/shortage of employees means a lack of sufficient human resources to perform specific tasks. It is a difficulty in recruiting qualified staff for both managerial positions and operational employees operating machines. Unavailability/shortage of employees can cause overload for current employees. This increases the risk of production errors, that is, product defects or organizational errors. The limitation also affects production efficiency and order fulfillment times. In the long term, it may affect the company's strategy and development by limiting its ability to increase the number of orders and invest in new technologies.

Internal conflicts between departments in a company can be the result of many factors. They most often arise as a result of poor communication, inappropriate division of responsibilities, lack of qualified staff, employee overload, and differing goals of individual departments. In many companies, individual departments strive to achieve different goals, often forgetting that there is only one goal: to make money today and in the future. Conflicts within a company affect the atmosphere and reduce work efficiency. In the printing industry, manufacturing a product in accordance with customer requirements requires the cooperation of many different departments. Conflicts can cause delays in order fulfillment, financial losses, or ultimately the loss of a customer. Eliminating conflicts requires actions such as clearly defining the company's goal, building an organizational culture, and dividing responsibilities and competencies between departments and their employees.

Employee incompetence means a lack of knowledge, qualifications, and experience necessary to perform duties in various job positions. In the printing industry, this includes knowledge of manufacturing processes and technology, the ability to operate specialized machinery, as well as effective management and decision-making skills. The incompetence of operational employees may result from a lack of adequate training in the first weeks of employment, as well as a lack of training to improve the qualifications of more experienced employees. The incompetence of middle and senior managers may be the result of a lack of sufficient knowledge about the specifics of printing processes, leadership skills, and general managerial knowledge, as well as being overburdened with responsibilities. Employee incompetence can lead to an increased risk of production errors, a higher number of complaints, material losses, and longer order fulfillment times.

Low employee engagement means a lack of motivation to work, a lack of responsibility for the tasks performed, and a lack of willingness to report and solve production problems. Low employee engagement can be influenced by salary levels, management style, or a lack of vision for employee development. In the printing industry, but not only there, timeliness and teamwork are important. A lack of engagement or a decline in engagement can lead to production errors, reduced product quality, downtime, and also affect the atmosphere at work.

In summary, each of the identified groups of constraints in the printing industry relates to a separate area of a company's operations. Operational constraints are mainly related to production. Market constraints mainly concern the company's environment. Organizational constraints are related to the internal structure of the company. Managerial constraints result from strategic decisions. Social constraints, on the other hand, mainly concern human resources.

The research conducted has shown that there are many factors limiting the development of the printing industry's production activities. Understanding these constraints allows for more effective identification and resolution of various types of problems that affect the functioning of enterprises. Properly identified constraints help a company focus its activities on the most important aspects and thus achieve its goals faster. Typical industry constraints can help a company realize what internal and external factors may hinder its operations and what actions should be taken to eliminate or minimize their impact. It is important not to treat this classification of constraints as a set of rigid rules, but as a tool for identifying and solving problems. Identifying the constraints of each industry should be treated as a dynamic process that requires constant updating and adaptation to changing market conditions and internal processes within the organization.

5.5 Company Size and the Occurrence of Constraints

It is quite interesting to examine whether there is a correlation between the size of a company and the occurrence of constraints hindering its development. The survey results obtained in the research process were subjected to statistical analysis. The Spearman's rank correlation coefficient R has been calculated to determine the relationship between company size and the possible constraints. The output is shown in Table 5.4.

The study showed that there is a statistically significant relationship between the size of a company (measured by the number of employees) and potential constraints that may arise in its operations. The R coefficients are statistically important for one of the pairs of variables. Statistical analysis showed a positive but weak correlation between company size and machine

TABLE 5.4

The correlation between company size and possible constraints

	Company size	
Constraint	***R* coefficient**	***p* < 0.05**
Machine capacity	0.143	0.0273

capacity constraints (R = 0.143). The result indicates that as the size of the company increases, so does the proportion of responses indicating insufficient machine capacity. It can therefore be assumed that this problem occurs more frequently in entities with a larger scale of operation and number of employees. The problem with machine capacity in larger companies is primarily due to a more diverse product range, larger scale of production, increased use of machinery, and a larger number of orders completed in a short period of time.

A rather interesting but also very important conclusion is that the other constraints considered in the study do not indicate any correlations. This means that regardless of its size, a company faces various constraints occurring in different areas within and outside. This suggests that operational, market, organizational, managerial, and social constraints can affect the development of small, medium, and large enterprises alike. There are no constraints that do not occur in companies of any size. Therefore, the developed group of constraints may be universal, that is, regardless of the size of the company, a constraint from each group may occur.

References

Ashlag, Y. (2015). *Zasady TOC: Zarządzanie ograniczeniami dla wzrostu biznesu*. MintBooks.

Goldratt, E. M. (1984). *The Goal*. North River Press.

6

Where Is Lost Value: Mapping of Value Stream and Bottleneck Identification

6.1 General Concept

This chapter focuses on increasing value and eliminating value losses within the operational system. As demonstrated above, value is wasted in an unnoticeable manner on a large scale during the daily operations of every business. Therefore, there is a need for analytical instruments support managers in the fundamental challenge of determining value and establishing strategies for eliminating various kinds of waste present in companies. Mapping value streams within organizations, along with identifying bottlenecks, offers highly effective strategy to address these issues.

The word value is being frequently used by many, including managers in organizations. However, its exact meaning in a particular organization's operation stream needs a real effort to be specific. When it is determined how the value is produced, it is possible to know where the value is lost and where the waste is produced in the process stream. The journey of preparing a map of value stream is about observation, tracking of passing material and work, gathering field data, in-depth understanding of the flow, and current limitations. When doing the mapping, usually many ideas on how to make the stream more efficient and having less waste come.

One of the proven mapping techniques is value stream mapping (VSM). It offers guidance for the preparation of a map along with a collection of symbols representing typical elements of a stream, which are the most important aspect to be determined while mapping. VSM is a good technique for mapping; however, it is not the map itself as a graphic image that is most important, but the journey of mapping, understanding value produced and waste lost, along with the resulting concepts on how to perform a stream in a better way. A business case on how VSM is employed is elaborated in this chapter.

Gathered process data, along with a comprehensive understanding of the whole stream, can work together with the concept of constraints, leading to the most effective improvements in the stream. However, bottlenecks

DOI: 10.1201/9781003655237-6

determination can be a real challenge in a real-world system of business operations. Using a single identification method out of many possible ones does not guarantee a reliable result or an unambiguous indication of the actual bottleneck. The reader will learn about a comprehensive methodology for identifying bottlenecks, comprising four stages and based on the use of several methods for identifying bottlenecks. Each stage of the multi-method framework for bottleneck identification is described in detail. At the end of this chapter, its practical application in the printing industry production system is presented. A step-by-step description of how to apply each stage of the methodology and its tools in a real production environment is provided. The results obtained are discussed and practical recommendations for its use are formulated.

6.2 Introduction

When an organization becomes aware of a gigantic scale of waste, it is then open and willing to look for tools that help determine in detail how this waste is created and support the search for new organizational formulas that provide more value and eliminate waste. The beginning of the improvement journey is always an in-depth understanding of value, sources of value, and where it is produced within the organization; next, the processes along with their cooperation and coordination system; and eventually where the value is wasted. The comprehensive view of the entire operation stream allows us to see the point that limits the flow.

One of the strategic improvement tools is VSM. It allows you to understand the flow of information and materials throughout the entire process, from the moment the customer places an order to the delivery of the finished product. It allows you to identify muda that hinder the realization of value for the product from the customer's point of view. The next strategic tool for improving value flow is the identification of bottlenecks. Effective identification of bottlenecks allows to recognize where flow is restricted, shorten production time, and increase the throughput of the entire system. There are many different methods that can be used to identify bottlenecks, but they all carry certain risks. Choosing the right method is not at all obvious or easy, as there is no guarantee that it will indicate the true result in the process. Correct identification of the constraint is the key to improving system throughput and taking improvement measures.

This chapter is devoted to techniques related to discovering the potential for improvements, along with the substantial concepts underlying these techniques. The mapping of the value stream is one of the primary techniques; the multi-method framework for bottleneck identification is another that directs improvements in the right way. The techniques and key concepts are explained in an approachable way, and the business case studies allow for understanding their practical application.

6.3 Value Stream and Its Mapping

Before the attention will be focused on the method for discovering where and how value is "produced" and where and how it is wasted, it is necessary to clarify what is a value stream itself, as this concept is noticeably different from others in this area, and in particular from the organizational processes. The value stream concept by its originality offers unique analysis frame along with precious insights.

The value stream is the entirety of activities carried out within from raw material and other inputs to the customer (Rother & Shook, 2009). This is a sequence of activities that are intended to create and then deliver value to the customer, or as modern marketing theory sees it, a value proposition. Therefore, the value stream encompasses all activities and the complete flow from one door, that is, the entrance, to the next door through which value exits toward the customer.

The value stream is therefore the physical path through the company from the first operation to the last—where the product is delivered to the customer. Along this value creation path, there will be both value-adding and non-value-adding activities; all of them are part of the value stream because the stream cannot be fragmented—it is a continuous sequence of mutually connected activities.

For value to flow, physical flows, work, and information are needed. The information flow runs from taking orders from customers to coordinating the shipment of products; the information flow ensures control and coordination of the flow at each of its stages. The physical flow involves the movement of input parts or raw materials through all processing phases until the finished products are delivered to recipients. Work flows occur along with and between the material flows.

The value stream integrates, in a single view, both the physical flows of materials and the flows of work, as well as the flow of information—all within a loop from external suppliers to the external customer. This horizontal and vertical consideration of all flows gives the value stream characteristics typical of an approach known as the big picture (Rother & Shook, 2009). The value stream perspective provides a complete view of all activities, their dependencies, and mutual interrelationships. This perspective allows one to see the entirety of complex interdependencies and avoid suboptimization when identifying improvements.

Value stream is not exactly what is meant by processes in the business process management (BPM) approach. For example, in the popular APQC classification of business processes, at the second level of the business process catalog, only 7 out of 59 listed standard processes involve direct contact with the customer, of course referring to the customer understood as the external recipient of the value created in the organization.

Thus, so-called process approach (BPM) assumes that a large portion of the processes within a company do not deliver an outcome of value to the external customer. In the process approach, regulatory processes and informational processes are analyzed separately as distinct groups of processes within the organization's process architecture. At the same time process approach pays equal attention to many processes covering other areas of the organization beyond the core sequence of operations that generate value.

Although the value stream does not directly focus on certain processes that serve as internal services to the main value creation flow, this does not mean they are overlooked. When such a service process impacts the dynamics of the value flow, it becomes a focus of interest during improvements design.

It is important to emphasize that in a typical business organization, and certainly in larger diversified companies, there are many value streams operating. There are therefore many sequences of activities and steering information that produce and deliver value. These sequences of activities are parallel but often intertwine and intersect with each other.

Identification and next in-depth study and diagnosis, followed by designing and restructuring the flow of the value creation stream, is the most general manner to describe what VSM is all about. It is a set of truly fundamental tasks that must lie at the absolute core of every organizational improvement methodology. Behind this is a very simple and incredibly powerful argument—the stream is responsible for generating money; if you don't focus on it, you will never achieve tangible results from your process improvement.

VSM is the most widely discussed mapping technique among researchers and consultants. VSM is a proven analytical and conceptual method thoroughly described by Rother and Shook (2009) in their notable book "Learning to See." This title aptly captures the essence of the VSM concept. Indeed, the VSM aims to help reveal something that managers do not always fully see—where value is created and how it is wasted.

The VSM method has also been published as an international ISO standard (ISO 22468:2020). According to the international standard, VSM is an effective tool for collecting, evaluating, and continuously improving the flow of products and information within an organization. The VSM methodology includes the analysis, design, and planning of value streams. Value streams in various sectors and composed of different types of processes should, according to the mentioned standard, be holistically improved based on the procedure defined by the standard (International…, 2020).

However, it is not just the technical skills of preparing a map, and in the case of the VSM method, drawing it, that are most important. Proper mapping requires skills in deeply understanding the value generated in the stream, how it is created, and the large amount of waste that arises simultaneously. Above all, conceptual knowledge of how to permanently improve, that is, streamline the processes, is crucial here.

Before starting to prepare a value stream map, it is important to clearly define what will be mapped. Typically, each organization delivers many products, and individual products differ from one another to a greater or lesser extent, requiring different processes for their creation and delivery. Therefore, it is essential to carefully specify which exact stream the mapping will concern. Usually, it is not possible to create a single map for the entire organization. You need to select a family of products that are similar in terms of the processes involved and map that stream. Later, you can map additional streams within the organization.

Creating a map involves tracing the product's path from the consumer back to the supplier and carefully drawing, using standardized symbols, each work process performed by machines and people, the materials, and the flow of information. One should ask how the future ideal flow might look. The authors emphasize that this should be done continuously and repeatedly, as it is the simplest and at the same time the best way to help oneself and coworkers understand what value is and what the sources of waste are (Rother & Shook, 2009).

As has already been emphasized above, mapping is not just drawing. Mapping is primarily about observing, asking questions to operators and to yourself, collecting data, determining where the value is, and understanding the specifics and dynamics of the value stream flow. Therefore, a good practice is to start this journey from the end, from the point where the stream meets the customer, in order to analyze all the steps of the stream in that light.

When moving upstream while mapping, the three layers must be thoroughly observed and understood (King & King, 2015, p. 5): the physical flow of material, the flow of all information, and the time elapsed in connection with each activity. First, attention is focused on the flow of material through halls, warehouses, areas, and work-in-progress fields; then special attention is given to the flow through processing operations (machines, workstations), always starting observations from the point where products are handed over to customers. All places where material or other inputs to the stream undergo processing must be examined in detail due to the parameters characterizing these phases.

At the same time, all types of information transmitted through various channels that control what is to be executed (produced) and when it is to be executed must be identified. The information flow starts from customer orders, is included in the schedules, and is in production orders, workstation plans, production instructions, control signals, material ordering, etc. Information usually flows in the opposite direction to the flow of work and materials.

The concept of a stream is inherently connected to its flow, and flow is always related to time. When mapping, the passage of time must be accounted for in all characteristic stages of the stream. It is necessary to determine the time associated with adding value as well as not adding value during the flow of the product.

During mapping, that is, examining the streams in a direct contact with it, it is necessary to determine all the parameters that define the flow dynamics. Attention should be paid to variables that may be important when redesigning it toward better value flow. Comprehensive information related to every processing step occurring in the stream must be collected.

There are many such measures characterizing processing operations in the context of value stream flow. The process cycle time (C/T) refers to the time spent processing one unit in a given process. Another important process parameter is the process changeover time, abbreviated as C/O. This refers to the period during which a given process (machine) is taken out of productive work in order to adapt it for processing a different product. Changeover breaks have a significant impact on the smoothness of the value stream flow.

Among the information of fundamental importance is also the scale of human resources involved in the process, that is, the number of operators performing the process. Another is the current availability time of the process, such as shift patterns and the number of work shifts during which the process operates. Quality in the process is also of basic importance for the smooth flow of the stream. Defects occurring in the processes cause the flow to slow down and result in higher work-in-progress inventory levels.

Another parameter is production capacity, which is the maximum amount of production that can be completed by a given process within a unit of production time. A parameter related to production capacity is the utilization rate of the process's production capacity, which is expressed as a percentage. Regarding inventory, it is necessary to precisely determine its scale at the various stages of the flow.

While mapping, all sources and helpful techniques should be utilized. It is important to remember the many available data sources, such as data recorded by digital interfaces of machines and devices, or from IT systems supporting production system management. Where this information is insufficient, measurements should be conducted, for example, time measurement using a stopwatch and counting the inventory levels between workstations.

The 5Why and 5W2H techniques can be useful, as they serve to deepen the understanding of peculiar issues faced in the stream. The 5Why technique aims to delve into the problem and avoid a superficial, and thus incorrect, understanding of it. It involves asking the question "why?" five times in succession regarding the problem being examined. The 5W2H technique with regards to mapping can serve as a checklist to ensure that the individual problems identified during the mapping process are thoroughly understood.

This method involves answering the following seven questions: what?, who?, where?, when?, why?, how?, and how much?

Along with a deeper understanding of the flow, a graphical form of the value stream map is created. Individual elements that make up the flow are marked on it using appropriate symbols. In VSM approach, the upper part of the map contains the flow controls, while the physical flows are placed below. At the very bottom, a chart is placed that compares value-added and non-value-added times. The time chart allows for the calculation of a basic indicator for the value stream: the value-added time ratio.

The VSM guide developed by Rother and Shook (2009) includes a set of conventional symbols that can be used when drawing a value stream map. However, it must be emphasized again that VSM cannot be reduced to the technical task of graphically representing the stream in the form of a map. Mapping is primarily a cross-sectional, quantitative identification and understanding of the flow.

Nevertheless, the graphical aspect—that is, the actual representation of the value stream—can be done in various ways, such as using a flowchart diagram technique. There are also organizations that use only tabular summaries listing all processes and basic information characterizing them as "a map" of the value stream. It should be noted, however, that a strong point of VSM is gathering and visualizing all the necessary information about the stream.

The most important phase of mapping begins when the value stream is fully understood and modeled. When its dynamics are completely identified, the way value is created is precisely defined, and where different types of resource waste occur is known. At that point, work can begin on developing a future state map, which will serve as a guide for implementing improvements in the stream. Even when mapping the current state, numerous ideas usually emerge on how to improve the flow of the stream.

It is advisable to use the same graphic technique for creating the future state map as was used for mapping the current state of the stream, except that the map presents the state of the stream incorporating significant improvements in its flow—the future state. Developing the future state map is a creative, design-oriented task aimed at ensuring better flow that will generate less waste. Proposed solutions should be ambitious but not impossible. They do not have to be immediately feasible; the map can include challenges for which detailed solutions will be sought through Kaizen improvement methods.

Among many possible VSM techniques, VSM appears to be the most comprehensive and convenient one. It provides a clear picture of all flow elements, it shows both value-adding and non-value-adding operations. At the same time, it effectively integrates all layers of the value stream. And most importantly, it allows for the proper preparation of a concept for an improved future state of the value stream that the organization will strive to achieve.

6.4 From Observation to Improvements—A Business Case Study

A small company employing nearly 100 employees is engaged in the production of metal garden furniture, such as tables, chairs, fireplaces, and yard decorations. The company uses simple manual technologies, including cutting, bending, joining, and painting; the primary material processed is black steel. This case serves as a good example for demonstrating VSM, as it consists of easy-to-understand production processes based on low-tech methods. The value stream of a metal outdoor shelf product is studied and summarized below.

Customer orders are not identical; however, their quantities are always a few hundred units. One of the orders was mapped for 387 shelves, which was placed three months in advance of the process start point. The material for the production of shelves was received into the warehouse two days before the start of the first production operation. The lead time (LT) for the order of the basic material, sheets of black steel plate, was 16 days.

The process of quality control and packaging is also responsible for order's shipment to customer. Quality control and packaging is performed in one place—a packaging area, both activities are carried out in one sequence, directly before starting packaging each item is carefully examined visually for its quality. 100% quality control is carried out. Quality control and packaging into individual packages takes 12–15 minutes (C/T) and is performed by a two-person team. Bulk packaging takes place on pallets, each containing 14 pieces. Data from the written quality control report for one shift are presented in Table 6.1.

The next upstream process is painting. The process consists of several steps that function as a sequence. First, washing is done in batches of four pieces, with a washing time of three minutes. After washing, the product dries on pallets for 0.5 hour. Then, in batches of 24 pieces are (1) painted and (2) heated, along with cooling, with each process lasting 1 hour. The LT for the entire sequence of processes is 168 minutes; however, the value-added time (C/T) for a single item is 96 minutes.

Polishing is the next process as an internal supplier of abovementioned painting process. Four single-person workstations perform this process. Batches of ten pieces, each on a pallet, are delivered to the workstation and then passed on to painting. The productivity statistics of the process are presented in Table 6.2. The cycle time for the polishing process is C/T = 7 minutes 48 seconds.

The field data obtained for the polishing process (Table 6.2) allows for a deeper understanding of how the process actually works. Aside from many detailed observations, it should be emphasized that the data clearly show very high variability in the process, even for the manual process performed in small batches and on a limited scale.

TABLE 6.1

Extract from the quality control report

	Values
Number of inspected	124
Number of returned to production items	3
Percentage of inspected (Qd, disqualified)	2.42%
Number of defects classified as minor and passed	8
Percentage (Qm, minor)	6.45%

TABLE 6.2

Statistics based on field data referred to polishing

	Item/hr
Mean	7.7
Median	7.5
Max	12.0
Min	4.3
Range	7.7
1/2 Range [as % of median]	51%
St. Deviation	2.4
Dev. [as % average]	31%
Average for the first 8 batches	6.7
Average for the last 7 batches	8.8

The welding of the shelf takes place at three workstations. The production batch size is ten pieces, and the batch fits on one pallet. The process product inventory amounts to 30 pallets. Basic productivity statistics were collected and comprehensively analyzed as above, similarly for all next processes. The process cycle time is C/T = 10 minutes and 10 seconds.

Bending large blanks is the next process. One operator is responsible for carrying out this process. Bending is done on a press brake. The machine setup time is C/O = 40 minutes. The production batch, understood as an internal order for the workstation, consists of 100 blanks, which fit on one pallet. The process cycle time is C/T = 41 seconds.

The first manufacturing process of the investigated stream is the cutting small blanks. The process is performed by two workers on a guillotine shear. Production orders for the workstations are 100 pieces each, forming production batches. The setup time is C/O = 30 minutes. Based on the gathered data, the process cycle time was determined to be C/T = 45 seconds. The VSM map is presented in Figure 6.1.

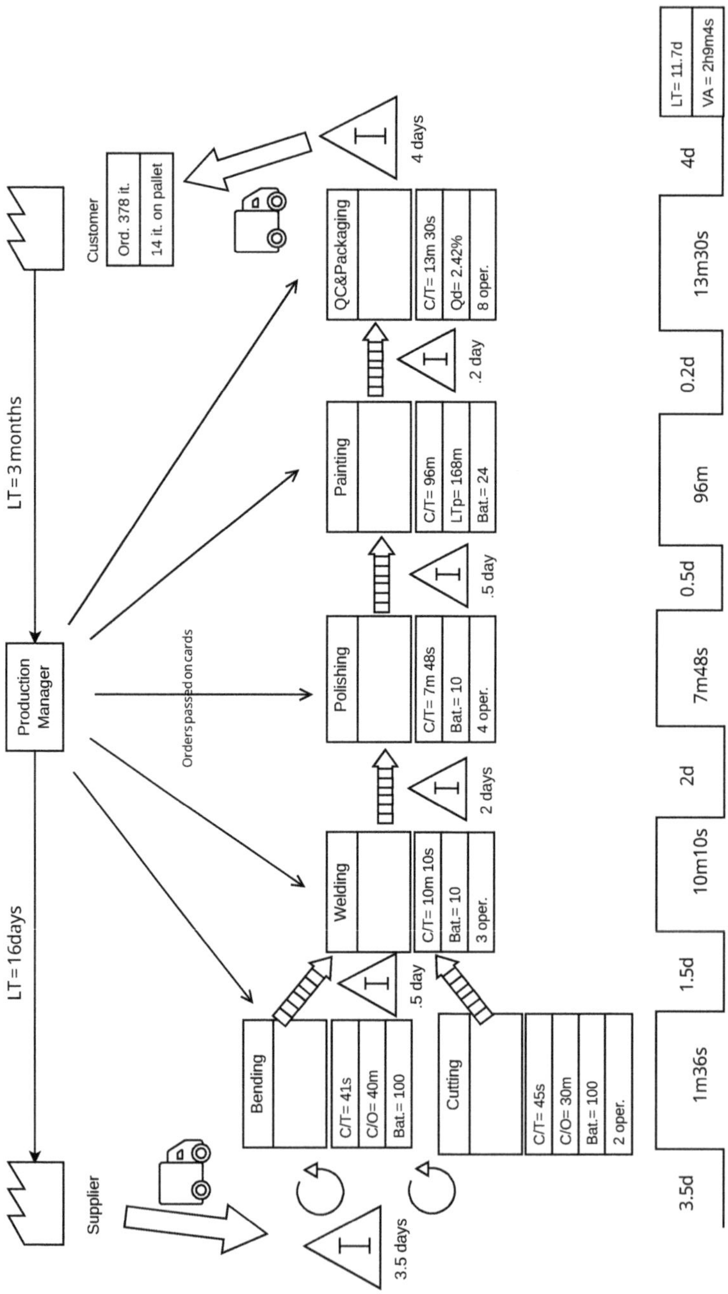

FIGURE 6.1
Value stream map of a garden shelf product.

The total time during which value is added is 2 hours, 9 minutes, and 4 seconds, so the CT/LT ratio is 0.008 (0.8%). It is clear how slowly value is being created in the stream. However, the mapping process allowed to discover many important improvements that the company might want to make and a future state map has also been developed. The three urgent meta improvements are as follows:

1. The processes are characterized by very high variability; it is necessary to develop and implement clear and effective work standards that enable predictable control and coordination. This will allow building a steady flow rhythm for the entire stream.
2. Currently, the stream places emphasis on final inspection, and the quality indicators are at a very low level. Steps should be taken toward control based on random sampling according to ISO 2859. Clear standards for self-inspection at each process must be established. It is essential to work on fostering responsibility for the work performed among its executors.
3. There is a need to build a system coordinated according to one process of the whole stream, which may be difficult given the current variability of processes. However, as these processes stabilize and begin to yield results, coordination should be planned in conjunction with visual inspection and based on a control process that sets the pace for the other processes. Visual FIFO (First In, First Out, the basic principle of flow control) zones can be used to manage flow control.

The VSM technique allows for a comprehensive diagnosis of process flow and primarily helps identify waste along with its systemic roots. It provides very good framework for designing process improvements and their coordination.

6.5 Comprehensive Methodology for Identifying Bottlenecks

The Theory of Constraints (TOC) focuses on identifying the main constraining factor that most limits the throughput of the system. The TOC is distinguished by its simplicity. It gives the promise of high effectiveness, which may determine its attractiveness in the eyes of managers. It is clear that the starting point is understanding the basic principles and the entire TOC. However, applying it in a real production environment can be a real challenge.

Precise identification of bottlenecks in a production system can be challenging for several reasons. First, due to the complex and dynamic nature of processes. Each system consists of several or a dozen or so interrelated processes that can change dynamically over time depending on the number of orders, availability of raw materials, machine load, breakdowns, etc. An additional challenge may be the need to collect reliable data on the basis of which it will be possible to determine that a bottleneck actually exists. This activity is subject to a certain degree of inaccuracy and error. Furthermore, the limitation is not always obvious and visible. The limitation may result from incorrect work organization, procedures, or inadequate resource allocation.

Another challenge is the possibility of using various analytical or observational methods. Several of them are discussed in Section 4.5. However, deciding which method to choose in order to correctly identify bottlenecks in a production system can be difficult for managers. The use of a single selected method for identifying bottlenecks (among many) does not guarantee that it will indicate the true result in the process. In his study conducted in a real production environment, Urban (2019) pointed to the problem of identifying bottlenecks. The application of several selected bottleneck identification methods indicated different processes.

To meet managers' needs for simple tools to identify bottlenecks in the production system, a methodology for identifying constraints was developed. The methodology consists of four stages: (1) in-depth examination of the production system, (2) identification of bottleneck, (3) determination of bottlenecks and the distance to the next level of constraint, and (4) verification of the correctness of the diagnosis. In total, the methodology comprises eight steps (Figure 6.2).

The first step is the identification of processes forming the product flow (#1 in Figure 6.2). This step requires a thorough understanding of all processes in the production system and identification of all operations that are performed during them. It is important to find mutual connections between individual areas in the system. Therefore, it is worth conducting direct observation of the process, taking into account not only production operations, but also extending the observation to include, for example, logistics, warehousing, and information flow processes. The main objective of this stage is not only to arrange processes in a logical order, but above all to gain a thorough and in-depth understanding of the processes taking place and their mutual relationships. Undertaking such activities allows us to learn about and assess the current state of the system.

The next step is to observe the inventory status before the process (#2 in Figure 6.2), which involves analyzing the inventory that accumulates between individual processes and controlling its optimal level. The observation should be carried out by moving along the process line. The accumulation of inventory before one of the processes may be due to the fact that its efficiency is lower than that of the others. Frequent downtime and low or no inventory further down the process indicate that this process is unable

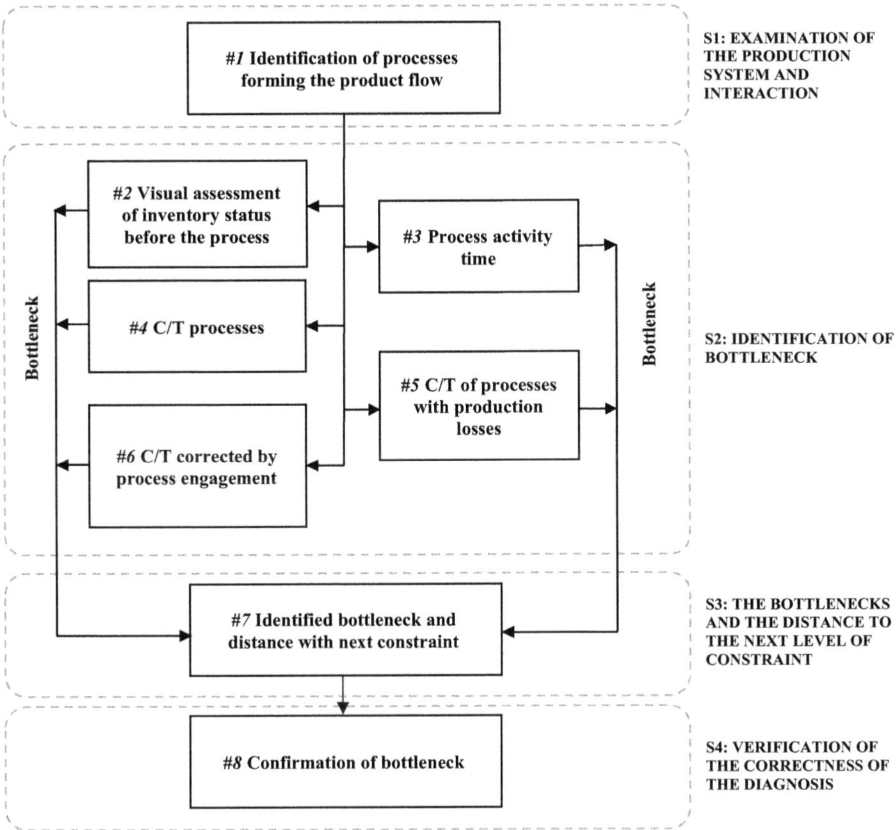

FIGURE 6.2
Multi-method framework for bottleneck identification.

to keep up with the pace of work and is a bottleneck. Regular monitoring of inventory levels is a simple way to detect bottlenecks.

Another method of identifying bottlenecks is to measure the activity time of individual processes and analyze them (#3 in Figure 6.2). It involves accurately measuring the total activity time of the machine/process in relation to the total available working time. Activity time is not only the time spent manufacturing products, but also the time spent on changeovers, downtime, adjustments, and other activities that affect the actual use of the machine. In relation to the total available operating time, the process with the longest time spent in the active state is considered to be the critical point of the system. Other processes with larger time reserves are unlikely to constitute a bottleneck. Depending on the data available to the company, activity time can be measured in two ways. The first option is to download data from the machine monitoring system. The second is to observe and directly measure the operating time of processes over a period that ensures its reliability.

Another suggestion of identifying bottlenecks is to use the process cycle time (C/T) method (#4 in Figure 6.2), which consists of determining how much time a given process needs to produce one unit of a product. The process that has the longest C/T compared to other processes limits the speed of the entire flow, that is, it is a bottleneck. Depending on the established product flow (#1 in Figure 6.2), C/T may also refer to loading or transport time. There are two ways to measure cycle time. The first is to measure C/T directly at the machine. The second is to estimate C/T based on data recorded by the company, such as the operating time of individual machines or the number of products manufactured in a given period. However, it is important to take into account differences in production conditions between the periods in which the measurements were taken and to consider any changes in the production process (seasonality of production).

C/T of processes with production losses (#5 in Figure 6.2) is another proposed step leading to the identification of bottlenecks in the production system. How should production losses be defined? At this stage, the production manager should take into account all factors that directly prolong batch processing, resulting, for example, from waiting for the availability of machines or processes, breakdowns, machine changeovers, or unplanned downtime that prevents or limits production at a given time. The time of losses should be determined using data recorded by the company. Including production losses in the C/T value of processes affects the final result. It determines how much time a given process needs to produce one unit of a product. The process with the longest C/T with losses should be considered a bottleneck.

The last of the proposed methods in the multi-method framework is C/T corrected by process engagement in production assortment (#6 in Figure 6.2). This method can be used where the system is not organized in a way that ensures a smooth flow of material (the product does not go through all stages of the production process). C/T-corrected method involves measuring C/T taking into account the actual load of processes resulting from participation in the production of a specific product range. As the name suggests, this method focuses on measuring the C/T of each identified process, but also on the degree of its involvement in the production flow. The product of these two indicators will indicate the actual time needed to manufacture the product in all identified processes. The adjusted C/T reflects the actual time needed to manufacture the product. According to this method, the process with the highest adjusted C/T is the bottleneck.

The next extremely important step is to identify the bottleneck based on the methods described above, and then determine its distance to the next constraint (#7 in Figure 6.2). What is the distance to the next constraint? It is the scale of the capacity shortage in relation to the next process, which is likely to occur after the current bottleneck has been eliminated. By determining the distance, it is possible to identify "how far" the next bottleneck is from the current one, which enables rational planning of measures aimed at increasing its flow. Obtaining such information provides a clue as to which improvement measures should be taken and which are not worth implementing.

According to Goldratt (1990), identifying a local constraint and taking improvement measures in that area will not significantly increase throughput. Identifying a bottleneck in a production system carries certain risks. It may happen that the identified bottleneck is not a real constraint of the system, and the effects of introducing changes based on incorrect assumptions will lead to "different" results. These may manifest themselves, for example, in the form of decreased efficiency, reduced production quality, no changes in the system, or simply a loss of valuable manager time.

In order to ensure the correct implementation of the TOC and fulfillment of the described conditions, the final step of the constraint identification methodology (#8 in Figure 6.2) has been proposed—confirmation of the bottleneck. This step consists of several questions designed to provide the manager with confirmation that a real constraint has been identified.

1. Have the processes that create product flow been identified and their interrelationships understood?
2. Is the identified process the weakest link in the production system that limits flow improvement?
3. Are the factors causing the bottleneck known?
4. Is the scale of the bottleneck's capacity shortage relative to the next constraint known?
5. Will the implementation of the proposed improvement measures lead to an increase in flow (increase in production volume)?

If all answers are positive, it means that the actual bottleneck in the production system has been located. Another constraint has also been identified, which is likely to occur after the first bottleneck has been eliminated. Acting in accordance with the TOC, the first step in improvement—identifying the constraint—has already been taken, and the manager can move on to the next steps, which consist of developing and implementing appropriate improvement measures to increase throughput. What if one or more answers are negative? The developed methodology does not require the use of all bottleneck identification methods, so one of the solutions aimed at obtaining all positive answers may be to use another method of identifying the constraint.

The effective implementation of the TOC in a company depends first and foremost on the correct identification of constraints—this is the basic step in continuous TOC improvement. Multi-method framework for bottleneck identification allows companies to organize their processes and provides a logical path for examining the production system. It enables a better understanding of the system as a whole and allows for a thorough understanding of its internal mechanisms. In the multi-method framework for bottleneck identification, each subsequent step is based on an increasingly broader view of the processes taking place and requires analysis of various additional information and the impact of factors. The manager does not have to intuitively search for bottlenecks, but can apply a structured and consistent methodology for identifying constraints. The manager uses clearly sequential steps and analyzes the process holistically, rather than selectively (as he or she deems appropriate at a given moment).

The multi-method framework for bottleneck identification consists of eight steps, not all of which need to be completed in order to effectively detect a bottleneck. While step #1 (in Figure 6.2) identified processes forming the product flow, step #7 identified bottleneck and distance with next constraint, and step #8 confirmed that bottlenecks are mandatory, each subsequent step can be applied selectively, depending on the available data and the specifics of the production system being analyzed. The identification of the constraint should be completed once it has been confirmed that the process in question is a bottleneck and the manager has obtained data enabling the distance to the next constraint to be determined.

6.6 Multi-method Framework for Bottleneck Identification—A Business Case Study

The subject of the study is a large printing company employing approximately 300 people. The company specializes in the production of printed packaging for the food and medical industries. Its product range includes labels, rigid packaging, and folding boxes. Production is carried out exclusively to customer order, and the company fulfills approximately 8,000 orders of various sizes annually.

S1: Examination of the Production System and Interaction

#1 Identification of Processes Forming the Product Flow

The production system under study consists of four basic processes, such as (1) offset printing, (2) die stamping, (3) die cutting, and (4) folding and gluing the product. Then, the finished product on pallets goes to the finished goods warehouse. The company also has a raw materials warehouse.

S2: Identification of Bottleneck

#2 Visual Assessment of Inventory Status Before the Process

In this step, the inventory for each process in the production system under study was analyzed. The number of pallets directly waiting in the production

TABLE 6.3

Inventory waits before each process

Process	Inventory waits for a process [pallets]
Offset printing	6
Die stamping	12
Die cutting	5
Folding with gluing	9

hall was counted. As part of the study, one observation was made during each of the two shifts in a single day, and the results are presented in Table 6.3.

The data indicates a high level of inventory prior to the pressing process. Observations show that this process works slower than other processes and requires frequent changeovers. Preliminary findings indicate that the method of observing inventory levels before the process shows that the die stamping process is a bottleneck in the production system under study. However, it is worth noting that these findings do not provide any indication of the relative scale of the bottleneck, that is, the scale of the capacity shortage in relation to the next bottleneck.

#3 Measurement of Process Activity Time

Data provided by the company was used to measure the activity time of each process. In order to obtain reliable results regarding bottleneck identification, a two-month period was selected during which the highest number of production orders was completed. At this stage, the number of machines involved in each process was also determined. It was determined that the offset printing process involved three machines—die stamping (three machines), die cutting (six machines), and folding with gluing (six machines). In the production system under study, the activity time included the operation of machines when manufacturing the product, machine changeover time, waiting time, and time spent cleaning the machines. The results of the analysis are presented in Table 6.4.

The results in the table indicate that the offset printing process has the longest activity time. It amounts to 902 hours. The measurement of process activity time method indicated that the bottleneck is the offset printing process.

#4 Measurement of Cycle Time of Processes (C/T)

In the printing industry, machines operate at very high speeds, which makes it impossible to measure cycle times directly at the machine. Therefore, data recorded by the company was used to determine cycle times. The data covered two months during which the largest number of customer orders were fulfilled. The calculations were based on the print run, that is, the number

TABLE 6.4

Measurement the activity time of each process

Process	Production time	Other activity	Total activity time per machine
		hr.	
Offset printing	966	1740	902
Die stamping	1670	985	885
Die cutting	2090	835	487
Folding with gluing	2070	1370	573

TABLE 6.5

Cycle times of processes

Process	Cycle time, C/T [sec/sheet]
Offset printing	0.31
Die stamping	**1.15**
Die cutting	0.65
Folding with gluing	0.71

of printed sheets produced in each process, and the time it took to produce that print run. In the folding and gluing process, the product is divided into several parts. This means that one printing sheet may contain several printed packages (multiplication of products). Therefore, in this process, the average number of packages printed on a printing sheet was calculated. The results of the analysis are presented in Table 6.5.

The results in Table 6.5 show that the C/T of each process differs from one another, which means that the efficiency of each process is different. The shortest C/T is for the offset printing process [0.31 sec/sheet]. This is the most efficient process among all those studied. The longest C/T is for the die stamping process [1.15 sec/sheet]. Therefore, it limits the flow of the system to the greatest extent and can be considered a bottleneck.

#5 Measurement of Cycle Time of Processes (C/T) with Production Losses

The application of this method to identify bottlenecks is more practical because it better reflects the actual conditions of the production environment. In every production system, there are many factors that will prolong the production of goods, that is, directly affect the entire production system.

Analysis of the company's data allowed us to identify seven types of production losses: products that do not meet quality standards, changeover time, machine cleaning, machine breakdowns, waiting for information, waiting

TABLE 6.6

Time of production losses

Types of production losses	Offset printing	Die stamping	Die cutting	Folding with gluing
	hr.			
Products that do not meet quality standards	10	0.5	2	15.5
Changeover time	1390	970	795	1200
Machine cleaning	260	10	40	170
Machine breakdowns	90	24	40	30
Waiting for information	56	-	-	-
Waiting for approval	36	0.5	0.5	0.5
Problems with input materials	79	0.5	0.5	-
Total	1921	1005.5	878	1416

TABLE 6.7

C/T with production loss time

Process	C/T with production loss time [sec/sheet]
Offset printing	0.85
Die stamping	1.80
Die cutting	0.9
Folding with gluing	1.2

for approval, and problems with input materials. The analysis period was two months. The results of the analysis are presented in Table 6.6.

The results presented in Table 6.6 indicate that not all production losses occur in every process. The loss referred to as waiting for information occurred only in the offset printing, and no problems with input material were identified in the folding with gluing. In the adopted two-month period, the time of production losses varied. The longest time occurred in the offset printing (1921 hours), and the shortest in the die cutting process (878 hours).

The next step is to determine the C/T of each process, taking into account the already determined production loss time. To this end, the production time and production loss time were added up for each process. The determined time was divided by the number of printing sheets produced in each process. The results are presented in Table 6.7.

The results presented in the table indicate that the C/T of processes with production losses differ from each other, and their values are higher than the C/T obtained in step #4. The longest C/T with production losses occurs in

TABLE 6.8

C/T corrected by process engagement

Process	Engagement [%]	Cycle time, C/T [sec/sheet]	Corrected C/T [sec/sheet]
Offset printing	100	0.31	0.31
Die stamping	60	1.15	0.69
Die cutting	100	0.65	0.65
Folding with gluing	90	0.71	0.64

die stamping (1.80 sec/sheet). According to this method, this is the bottleneck of the tested production system.

#6 C/T Corrected by Process Engagement

The printing industry is characterized by a diverse range of products, which involves the use of various production processes depending on customer requirements and the specifics of the order. The production system under study is organized as series production. Batches can be as large as 150,000 pieces. In this case, it is reasonable to use the adjusted C/T method based on process involvement.

Based on the data recorded by the company, the engagement of each process in the production of goods was determined. The amount of input was compared with the total input resulting from production orders. It was determined that offset printing and die cutting are fully engaged in the production of all manufactured products (100%), while die stamping is 60% engaged and folding with gluing is 90% engaged. Next, the basic C/T values of the processes calculated in step #4 are adjusted (multiplying) by their engagement. For example, die stamping is engaged in 60% of total production, which means that the adjusted C/T is 0.69 [sec/sheet] with an actual C/T of 1.15 [sec/sheet]. The results are shown in Table 6.8.

C/T corrected symbolically reflects the time needed to manufacture a product that serially goes through all processes. Analysis of the data in the table shows that the highest value was obtained for the die stamping process, so it should be considered that, according to the method used, this is the bottleneck of the production system under study.

S3: The Bottlenecks and the Distance to the Next Level of Constraint

#7 Identified Bottleneck and Distance with Next Constraint

All methods were used to identify the bottleneck. A total of five methods were used, and the results of the analysis are presented in Table 6.9.

The case study presented showed that identifying bottleneck is not easy. Depending on the method of analysis used, bottleneck can appear in

TABLE 6.9

Identification of bottlenecks—results of the methods used

Method	Bottleneck
#2	die stamping
#3	offset printing
#4	die stamping
#5	die stamping
#6	die stamping

different places. The process activity time measurement method (#3) indicated that offset printing was the bottleneck. The other methods indicated that die stamping was the bottleneck.

The case described indicates that if processes are analyzed according to the activity criterion, the bottleneck emerges in offset printing, while if the analysis is based on an increasingly broader view of the processes, taking into account factors such as production losses or the degree of engagement, then the bottleneck appears in a different place.

It is worth noting that the bottleneck is not always solely due to the machine's operating time or load, but can be influenced by a number of other factors such as changeover time, quality losses, and degree of process engagement. Therefore, it is important that managers do not limit themselves to a single method of identifying bottlenecks. The results of different approaches should be compiled and compared with the actual functioning of the production system. Taking such actions makes it possible to reliably determine which process actually determines the throughput of the company and should be the subject of improvement measures.

Once it is known which process is the bottleneck—die stamping—the next step is to determine the distance to the next constraint (the next level of constraint). The distance to the next constraint is the scale of inefficiency to the next constraint. Therefore, it will be the process that is likely to emerge as the next bottleneck after the first one is removed.

The corrected C/T results presented in Table 6.8 indicate that the next bottleneck will be the die cutting process (corrected C/T is 0.65 sec/sheet) and, closely behind, the folding with gluing process (corrected C/T is 0.64 sec/sheet). What is the scale of the productivity gap between the current bottleneck and the next one, which is likely to emerge after it is removed? The hourly productivity of the current bottleneck is approximately 5200 sheets/hr, while that of the next bottleneck, that is, the die cutting process, is approximately 5500 sheets/hr, and that of the folding with gluing process is approximately 5600 sheets/hr. The distance to the next constraint is approximately 300–400 sheets/hr. This information is important for future planning of how to remove the current bottleneck.

S4: Verification of the Correctness of the Diagnosis

#8 Confirmation of Bottleneck

The final stage of the methodology consists of five questions designed to provide the manager with an answer as to whether a real constraint in the production system has been identified.

1. Have the processes that create product flow been identified and their interrelationships understood?
   Yes. All production processes have been thoroughly understood, as have their interrelationships.
2. Is the identified process the weakest link in the production system that limits flow improvement?
   Yes. The bottleneck identification methods used indicated that die stamping is the weakest point in the production system, which consequently limits flow improvement. Die stamping has the lowest hourly throughput, which is 5200 sheets/hr.
3. Are the factors causing the bottleneck known?
   Yes. Compared to other processes, die stamping has the longest cycle time of 1.15 sec/sheet, so it is time-consuming. In addition, the analysis of factors prolonging batch processing is as high as 1000.5 hours. The largest share is changeover time, amounting to 970 hours over 2 months. Therefore, the manager should first take improvement measures aimed at reducing production downtime by shortening the die stamping changeover time.
4. Is the scale of the bottleneck's capacity shortage relative to the next constraint known?
   Yes. The scale of the capacity shortage relative to the next constraint is approximately 300–400 sheets/hr.
5. Will the implementation of the proposed improvement measures lead to an increase in flow (increase in production volume)?
   Yes. Shortening the changeover time will increase machine availability, that is, the time available for product manufacturing. This will increase throughput, which will enable an increase in production volume.

Providing all positive answers from the proposed list of questions indicates that the actual bottleneck in the production system has been located. In addition, another constraint has been identified, which would likely emerge after the first constraint has been removed.

From a manager's perspective, all the proposed stages of the multi-method framework for bottleneck identification are practical and can be directly applied in the production system. The value of the methodology lies in its broad, comprehensive view of processes, which is not limited to the analysis of a single, selected area, but covers a wider scope.

Two types of methods were used in the multi-method framework for bottleneck identification. The first type of method is observation of the production system (#2). Observation should be thorough and take into account a deep understanding of the production flow. The causes of situations should be taken into account, not just their effects. It is extremely important to make several observations at different times, for example, during different production shifts, in order to obtain more reliable results and not just momentary fluctuations in production.

The second type of bottleneck identification method uses data recorded and collected by the company (#3, #4, #5, #6). Collecting such data is necessary for the proper analysis of production processes and subsequent identification of bottlenecks. However, in order to obtain reliable results, it is necessary to select a period of time that is representative of the production under study, that is, one in which the company carried out a large number of orders. In the industrial sector, this period is often referred to as production seasonality. This will ensure that the results of the analysis are reliable and provide a better understanding of the processes, which in turn will contribute to more effective identification of bottlenecks.

The final step in the multi-method framework for bottleneck identification is based on determining the distance to the next constraint. The method of identifying bottlenecks based on a visual assessment of inventory levels prior to the process does not provide information on the scale of the capacity shortage in comparison to the next process. It is difficult to determine the size of the bottleneck in the context of subsequent processes. Therefore, it is recommended to supplement such an analysis with an additional method that will allow the scale of this shortage to be determined. In this way, the manager will have a more comprehensive picture of the functioning of the production system, which will allow for better decision-making regarding its optimization and improvement.

When identifying bottlenecks in the production system, it is worth using the C/T measurement method (#4), that is, the time needed to produce one unit of the product. Each subsequent method of identifying constraints is based on an increasingly broader view of the processes, but ultimately refers back to the basic C/T value. The information needed to apply each subsequent method will only supplement it.

The multi-method framework for bottleneck identification does not require the use of all methods for searching for bottlenecks. However, managers must be vigilant, as the use of a single method can lead to erroneous results. Therefore, it is recommended to use at least two methods for identifying bottlenecks based on different parameters of the production system in order to reduce the risk of misdiagnosis. The five questions stage (#8) should not be omitted, as it will help you decide whether a real constraint in the production system has been identified.

References

Goldratt, E. M. (1990). *What Is This Thing Called the Theory of Constraints and How Should It Be Implemented?* North River Press.

International Organization for Standardization. (2020). *ISO 22468:2020, Value Stream Management (VSM)*. ISO.

King, P. L., & King, J. S. (2015). *Value Stream Mapping for the Process Industries*. CRC Press.

Rother, M., & Shook, J. (2009). *Learning to See: VSM to Create Value and Eliminate Muda*. Lean Enterprise Institute.

Urban, W. (2019). TOC implementation in a medium-scale manufacturing system with diverse product routing. *Production & Manufacturing Research*, 7(1), 178–194.

7

Improvement: Kaizen, DMAIC Cycle, and Five Focusing Steps

7.1 General Concept

The basis of every organization's strategy is the process of continuous improvement. There are many different tools in the literature that relate to process optimization. Among the many, three approaches are worth noting, namely, Kaizen, DMAIC (Define, Measure, Analyze, Improve, Control), and Five Focusing Steps (5FS). Each of them derives from a different methodology, but all aim to improve the organization.

Lean Management refers to the Kaizen culture, which is its foundation. Kaizen is the search for opportunities to eliminate waste and deficiencies in production and organizational processes. These are mainly small improvements proposed by company employees, usually in the area where they work. DMAIC is a Six Sigma tool consisting of five stages (Define, Measure, Analyze, Improve, Control). It allows you to determine how processes function in the company and, based on data and facts, implement improvements in the problem area. The last one, 5FS, is a tool derived from the Theory of Constraints (TOC). It focuses on identifying system constraints and eliminating them, thereby increasing the throughput of the entire system.

An analysis of selected tools has shown that despite differences in their operation, all three aim to improve the organization. The key to success is not choosing a single tool or improvement technique, but the ability to consciously integrate them. Kaizen, DMAIC, and 5FS have great potential to complement each other, and integrating these methods can lead to synergy.

7.2 Introduction

Companies operate in a constantly changing and highly competitive environment and face high customer demands. In such circumstances, a company's

DOI: 10.1201/9781003655237-7

ability to continuously improve and effectively manage its production system is key to its success and development, as well as to maintaining its position in a competitive market.

Managers facing these challenges can use proven process improvement tools. Three tools stand out: Kaizen, DMAIC, and 5FS. Each offers a different, unique perspective on the improvement process. Kaizen is a solution for smaller-scale improvements based on employee involvement. However, not all problems in a company can be solved with simple tools. There are situations where tools with a broader perspective are required, such as DMAIC or 5FS.

This chapter presents the basics of Kaizen, DMAIC, and 5FS. Their practical application is outlined, with particular attention paid to the types of projects in which they can be used and bring the greatest benefits. A key element of this chapter is the identification of synergies between these tools, taking into account their potential to complement each other.

7.3 Integration of Process Improvement Methodologies

In the last few decades, there has been interest in integrating Lean Management, Six Sigma, and TOC, when business practitioners and scientists noticed certain limitations when using them individually in production systems. Despite the numerous successes that companies can achieve by using each of them, it can be seen that process improvement based on a single methodology can bring strong but limited results. Contemporary systems are too complex to apply a single, isolated approach. The fact that companies use more than one improvement methodology at the same time has been demonstrated in the research described in Chapter 3.

As early as the 1990s, publications appeared on the possibility of combining Lean Management with the TOC (Goldratt, 1990), in which it is mentioned that improvements in one area can worsen results from the point of view of the entire company, and Lean Management tools that allow for streamlining the flow only make sense to use once a constraint has been identified. In the following years, interest in the integration of Lean Management and Six Sigma also grew (George, 2002; Basu, 2009), combining the elimination of non-value-added activities with the reduction of process variability, resulting in a methodology known as Lean Six Sigma (LSS). Subsequently, these three methodologies, Lean Management, Six Sigma, and TOC, were also combined to create a model called TLS (Nave, 2002; Kose & Agdeniz, 2018).

In today's business environment, it is extremely important to skillfully and consciously integrate different approaches to process improvement and the tools they offer in such a way that they complement each other and do not interfere with each other. The most interesting and promising direction in this area is the integration of LSS and TOC tools.

Each of these methodologies is characterized by an original approach and the principles and tools that prevail in them. In general, Lean Management focuses on improving value stream flow, eliminating muda, and determining added value from the customer's perspective. Six Sigma focuses on reducing process variability and reducing the risk of errors based on data, and one of the methods of conducting a project is the DMAIC cycle. The TOC indicates that every company has constraints that hinder its development. Identifying these constraints and focusing the efforts of the entire organization on eliminating them can bring measurable benefits.

Each methodology has its own strong and unique advantages. However, the synergy of these methodologies' tools allows for a better view of the company as a single organism—that is, a system in which local improvement gives way to global optimization. The integration of LSS and TOC tools is an attempt to answer the question: How can processes be improved globally, in a way that benefits the entire system? This is also one of the key objectives of this book, which aims to show how to combine different tools to support systemic and sustainable improvements.

7.4 Process Improvement Tools

Separately, all three methodologies clearly indicate that organizational culture is an integral part of the continuous improvement process. Organizational culture of improvement refers to the conditions, principles, values, etc. that are characteristic of a given organization. It determines whether an organization is capable of introducing changes and then maintaining them. Without an organizational culture of improvement, all possible methodologies, techniques, and tools will only result in temporary changes in the company, with no prospect of survival or measurable benefits.

Of the three, Lean Management makes the greatest contribution to the area of organizational improvement culture. This methodology is most closely related to the Kaizen culture, which is its foundation. Kaizen is most often associated with the continuous improvement of activities, processes, and resources.

Kaizen was first used in Toyota factories. However, its roots lie in Japanese culture, reflecting Eastern thinking and lifestyle. Kaizen itself means improvement, that is, continuous improvement of the current state. It is not just another project that has a beginning and an end. Kaizen is a way of thinking for every employee. Its strength lies in the small, everyday improvements proposed by the company's employees. It focuses on the constant search for opportunities to eliminate waste and deficiencies in production and organizational processes (Hamrol, 2018, p. 95).

Kaizen is based on the belief that employees are the best people to identify problems and propose solutions. Kaizen ideas can be submitted by any employee, regardless of their level, from an operational employee to the company president. Kaizen supports Lean Management mainly in eliminating all forms of muda (Van Kollenburg, 2021) described in Chapter 4.

Kaizen can be implemented in companies in various forms. The first possible form of improvement is the so-called Kaizen event. It is a several-day event during which teams focus on a selected area or process in order to identify problems and solve them immediately (Franken et al., 2024; Vo et al., 2019). Another form is the so-called daily Kaizen. It consists of small, daily improvements made by employees at every level of the organization who are involved in improving their workplaces and processes. The goal of daily Kaizen is to eliminate waste (Imai, 2007, p.15; Owieczka, 2024, p. 38).

One of the most commonly used tools supporting Kaizen is the employee suggestion box. It is a system for submitting employee ideas for improvements, often related to their workplace. In its traditional form, it can be a physical box into which employees place cards with their improvement ideas. In other companies, it can take a digital form, such as an online platform or application.

Kaizen is characterized by a fairly rapid and high rate of problem-solving implementation. The implementation process takes place in a short period of time, as decisions can be made at the immediate supervisor level within a few days of the report. Employees are encouraged to report even minor suggestions for improvement. Kaizen in the workplace builds a culture of engagement within the company and shows that ideas submitted by employees are analyzed and taken seriously (Van Kollenburg, 2021).

One of the tools for solving problems in companies and supporting Kaizen is the A3 Report. It is a simple tool that presents the essence of the entire problem and ideas for its solution on a single sheet of A3 paper (hence the name of the tool). The A3 Report can be used, among other things, to report improvements in the company, but also to communicate improvements to others. The power of this tool lies in its simplicity. A typical A3 Report consists of a description of the problem, a presentation of the current state, an analysis of the root causes, proposed solutions to the problem, and an implementation plan.

Kaizen eliminates everyday problems based on employee involvement. However, it is a good solution for smaller-scale improvements. Not all problems that arise in organizations can be solved with simple tools. More complex problems in processes require methods with a broader perspective. In such situations, the DMAIC cycle, one of the basic Six Sigma tools, plays a crucial role.

The Six Sigma methodology is a data-driven, statistical approach to reducing process variability, thereby minimizing errors and defects in organizational processes. The overarching goal of this methodology is to increase customer satisfaction and reduce costs associated with poor quality.

In Six Sigma, the DMAIC cycle is a systematic and repeatable approach to improvement. It is not a random sequence of steps. DMAIC is a carefully planned five-step process, each step building on the previous one. DMAIC guides the team from the moment a problem is identified to the measured and verified improvement of that problem.

DMAIC forces a certain way of thinking and conducting improvement activities. You cannot move on to the next phase until the previous stage has been reliably completed. This sequence of steps makes DMAIC a unique and valuable tool that works well in situations where intuition fails and random ideas for solving problems lead to temporary relief without permanently eliminating the cause (Six Sigma, 2018).

DMAIC in a company is project-based. It lasts from several weeks to several months and is led by a team of people, often certified Green Belts or Black Belts. Such a project has a clearly defined goal and scope, financial objectives, and a predetermined formal oversight system, that is, a system of accountability, decision-making, and project progress monitoring. The DMAIC cycle is used when a company has a problem of high business importance to solve, where reducing process variability can bring operational as well as financial benefits.

The first phase of the DMAIC project is the Define phase. It is the foundation of the project, on which the entire improvement process is built. It involves identifying the problem and determining why it needs to be eliminated and what results the project is expected to deliver. A well-executed Define phase allows you to answer several key questions: What is the problem? Are customer requirements really not being met? What are the customer's requirements? What are the boundaries of the project? What financial and operational benefits will it bring?

At this stage, it is important to determine whether the process actually deviates from the customer's requirements/expectations. Therefore, an honest conversation with the customer should be conducted in order to learn their Voice of the Customer (VOC) and define Critical to Quality (CTQ), that is, critical quality parameters that can be measured. Once the customer's requirements are known, the next step is to learn about the process, its boundaries, and understand the overall flow, using, for example, a SIPOC map.

In the Define phase, it is important to establish a project team and assign roles. The project team should consist of people performing various functions within the company. The causes of a problem rarely lie within the scope of a single department. Therefore, in addition to people directly involved in the process, the team should include people who provide inputs to the process and receive outputs. A kind of summary of the Define phase is the creation of a Project Charter with defined measurable project outcomes. It is a kind of contract between the Green Belt accepting the task and the Sponsor who assigns the task (Eckes, 2016).

One of the most important foundations for the subsequent stages of the Six Sigma project is the Measure phase. It is a kind of transition from employee intuition and assumptions to hard data and facts. Without a well-conducted Measure phase, there is no certainty about the results obtained, that is, whether the identified causes are real and whether the proposed solutions could bring about a reduction in process variability.

In the Measure phase, it is important to achieve three key objectives. First, identify the key process indicators that relate to the defined problem and CTQ. Next, ensure that the measurement systems are reliable and credible by using Measurement System Analysis and, for example, the Gage R&R technique—Repeatability and Reproducibility study. The final objective is to collect current data from the process, on the basis of which final conclusions will be drawn. Without accurately collected and, above all, reliable data, a manager is unable to identify the real causes of problems or take effective corrective action. That is why it is so important that each stage of this phase is carried out carefully and thoroughly.

The next phase is Analyze, which involves searching for the root causes of identified problems using statistical tools. In this phase, the team must demonstrate a certain degree of humility. Since the cause determining process variability, defects, and other unfavorable phenomena in the process is unknown, it is necessary to trust the data and conclusions obtained. It often happens that intuition fails and the real cause of the problem lies elsewhere.

At the initial stage of the Analyze phase, it is important to establish as many hypotheses as possible that will relate to the cause of the problem occurring in the process. In other words, all possible factors that influence the phenomenon observed in the previous phase should be identified. At this stage, it is possible to use various tools that support systematic thinking and team creativity, such as the Ishikawa diagram, Pareto diagram, or 5 Whys analysis. In the next stage, the team undertakes empirical verification of the hypotheses using data and statistical tools. Depending on the type of data, the team may use, for example, correlation analysis, linear regression, analysis of variance, *t*-tests, or nonparametric tests. The purpose of this stage is to find the causes that are actually responsible for the irregularities in the process (Pyzdek & Keller, 2010).

The next phase of the DMAIC cycle is Improve. This is the phase of the project in which the team seeks a solution or solutions to problems based on the root causes identified in Analyze. The result of the Improve phase is a proposal for a team action plan that will ultimately improve the process and its variability.

The solutions proposed by the team are implemented under controlled conditions. This means that the team first tests them on a limited scale, for example, on a single production line, measuring their impact on eliminating the problem. When the proposed solution is effective and eliminates the problem, full-scale implementation follows. If several solutions are proposed, they should be implemented one after the other, at intervals, to see if they produce the expected result. Implementing several solutions at once will not provide a clear answer as to which solution actually eliminated the previously identified problem.

Control is the final phase of the DMAIC cycle, which determines the success of the project. Its main goal is to ensure that the adopted solutions are stable, actually work in the company, and bring lasting benefits. This usually involves standardizing new procedures, training employees, implementing systems for monitoring process indicators, and developing a plan to respond to any deviations from the adopted standard. In the Control phase, the team develops a plan to prevent the process from reverting to its pre-improvement state (McCarty et al., 2004).

Both Kaizen and the DMAIC cycle focus on the continuous improvement of company processes. Kaizen usually deals with small problems and everyday ideas for improvement reported by employees, while DMAIC is a systematic improvement of process variability in problem areas. Both approaches assume that it is always worthwhile and possible to improve something. From the perspective of Goldratt's TOC, it is important to consider whether improving every problem area is actually justified. Or should the company focus its efforts only where it matters to its overall performance?

The TOC, like Kaizen and DMAIC, is characterized by a systematic approach to continuous improvement in a company. The difference is that the TOC focuses on identifying the most important constraint in a company, which at a given time most significantly limits the throughput of the entire system. The TOC is based on a systematic thought process called the 5FS. The 5FS is not definitely an improvement procedure. It is rather a managerial logic that helps to guide changes from an organization-wide perspective. As the name suggests, these are five steps performed one after the other.

The first step is to identify the constraint, that is, to pinpoint the location that most severely limits the system. The TOC suggests that improving a resource that is not the critical constraint will not affect the performance of the entire system. Identifying the key constraint of the company will indicate the direction for improving the synchronization of various areas of the enterprise. Goldratt and Cox (2004) argue that every company has at least one constraint. If nothing limited the company, it would achieve infinite value. Constraints can be found in various areas of the company (more on constraints and examples in Chapter 5). In addition to identifying them, it is important to rank them in terms of their impact on the achievement of the goal.

The second step is to exploit the constraint, that is, eliminate various factors that prolong work on the constraint and maximize its use, that is, get more out of the constraint. Every minute of constraint unproductivity is a loss for the entire system. Goldratt (1984) argues that there are large reserves of utilization in constraints, so once a constraint has been identified, its operation should be carefully examined. The analysis will reveal unproductive activities and those that prolong the duration of the constraint. Therefore, the unproductive time of the constraint should be minimized and its uninterrupted, continuous operation should be pursued. However, it is important that the actions taken in this step do not incur additional costs.

The third step is subordination, that is, synchronizing the pace of work of the other elements of the system with the constraint. In other words, it is adjusting the pace of slowing down production orders to the processing capacity of the constraint. In traditional production management, there is a belief that resources that are not working only generate losses. The effect of continuous machine operation can include overproduction, large inventories of work in progress, and queues that threaten flow. There is no need for resources that are not constraints to produce more, because the performance of the entire system is determined by its constraint anyway.

Step 4 is to strengthen the constraint. When all possibilities for increasing the throughput of the process have been exhausted and the constraint still has not been eliminated, further action must be taken to remove it. Step 4 allows for investing funds in the limiting factor, for example, in additional machinery, new technology, hiring an employee, or launching an additional production shift. At this stage, the constraint should be eliminated (Dettmer, 2001, p. 53). This step is not always required; it depends on the company's strategy (Ronen & Pass, 2001).

Step 5 is to restart the 5FS cycle, that is, return to Step 1. If the actions taken in the previous steps have produced the expected result of eliminating the constraint, then the constraint has automatically moved to another location. Therefore, the continuous improvement process must be restarted in order to search for the next constraint (Goldratt, 1984).

Kaizen, DMAIC, and 5FS are three selected tools for continuous improvement in organizations, which are linked by the PDCA (Plan-Do-Check-Act) cycle logic. However, each of them responds to different business needs. Kaizen refers to small, everyday, usually minor improvements. The DMAIC cycle is about reducing process variability and limiting the risk of errors. 5S, on the other hand, is a kind of strategic approach to optimizing the entire system, focusing on identifying the company's key constraint and eliminating it. This book aims to develop some guidelines for the coexistence of these three tools, highlighting how they can reinforce each other.

7.5 Simultaneous Use of Methods

The efficiency of a management system in an organization lies not in following one "golden" tool for improvement, but in the ability to consciously integrate several of them in the most favorable configuration. The methodologies presented in the previous section have great potential to complement each other, leading to several synergies.

It is worth seeking answers to the question of how to combine these approaches in practice. Of course, it is not about using them all at the same time, as this would lead to chaos and disorganization of the entire system. A clear framework for action is needed to integrate these methods. Below are some tips for combining the strengths of Kaizen, DMAIC, and the 5FS into a single, coherent system of process improvement, with each complementing the others.

Coordinated use of the three tools must start with constraint identification. It involves identifying the strategic constraint of the company that has the greatest impact on system throughput. Focusing improvement activities on the wrong areas will lead to a waste of company resources. Without identifying the key constraint of the system, the actions taken in the company will attempt to move toward steady improvement but with varying results.

For constraint identification, the multi-method framework for bottleneck identification described in detail in Chapter 6 can be used, along with key Lean Management tools such as Value Stream Mapping. This tool will show the flow from supplier to customer, paying particular attention to the flow of material and information throughout the system, work in process, process durations, and waiting times. When performing VSM, you should also identify the basic types of waste (more in Chapter 4).

Another indispensable task is to define value from the customer's perspective, which is the foundation of both Lean Management and Six Sigma projects, revealing this perspective as the VOC. The information obtained should be presented as CTQ, that is, measurable parameters critical to customer satisfaction. This will lead to determining the direction of improvement actions on the identified constraint that will actually bring value to the customer.

The TOC, along with its improvement routine 5FS, implies adopting one indicator for the whole system, such as throughput, which can serve as a navigator for process improvement actions. Other navigation indicators are also possible, but they must meet the requirements of a global company navigator. Such an indicator will help avoid local sub-optimizations and maintain focus on improvement projects that contribute to the results of the entire system. However, it is extremely important, while improving throughput, to remain focused on delivering value to the customer.

The identified company bottleneck needs to be examined more deeply to assess its scale and how it affects the entire system. This involves measuring current constraint indicators, such as cycle time, number of downtimes, changeover time, number of defects, and all other factors affecting the uninterrupted operation of the constraint. The current state of the constraint is determined before the improvement process begins (process baseline).

Measuring and carefully examining the bottleneck allows you to distinguish between subjective observations and actual factors affecting its performance. It also indicates how far the bottleneck is from the required level of performance. A company must identify which factors actually affect throughput and to what extent the bottleneck deviates from the desired level of performance.

A deep understanding of the bottleneck implies data analysis; the experiences of the DMAIC method would be particularly helpful. All available information and data have been explored to assess which factors have the greatest impact on working at full potential. Are these downtimes, changeover times, unpredictable variability, operational errors, or other flow disruptions?

Root cause analysis (e.g., 5WHY) and Lean-based observations will help identify the factors that affect throughput constraints. The TOC perspective allows you to evaluate them and determine their greatest impact on the flow of the entire system, not just a local part of the process. All this should lead to identifying the real causes of the constraint.

The potential of the constraint must be maximized before the company begins to invest in new improvement solutions, such as new resources or process expansion. Now, the full potential of DMAIC and Kaizen methods should be exploited. Projects focused on elevating the constraint and ensuring decreased variability along with increased predictability of the other processes in the product streams will yield real benefits in terms of global navigation indicator growth, as stated above.

Continuous waste elimination will ensure maximum flow through the bottleneck and maximize throughput. A stabilization phase is inherent in each of the three methods. Kaizen, like the one based on the A3 Report, culminates in the introduction of a new standard for a given section. From the perspective of the DMAIC cycle, this is the Control phase, in which indicators should be controlled, the process monitored, and work standards applied to prevent a return to old practices. Control provides a sense of predictability and stability to the process and also allows for early detection of certain deviations from the standard.

A huge advantage of Kaizen projects is that they rely on the engagement of ordinary operators. Employees have a significant impact on maintaining established standards, as well as on strengthening and further improving them. Employee attentiveness and commitment contribute to the early detection of problems and significantly affect process stability.

Integrated use of the three methods, along with the strengths of the methodologies behind them, offers managers instruments for a new perspective on organizational dynamics and the harmonious integration of managing organizational growth, business environment factors, and the human-social internal environment. The main synergy emerges from good harmonization and synchronization of change projects in processes within the whole organization.

Mutual support of methods and the different philosophies behind them for improving the organizational system is not a simple sum of "5FS + DMAIC + A3," but it leads to a new quality. Synergy, which involves maximizing the upward trajectory of the results curve, is what every manager strives for.

This combined approach can ensure the highest ROI (Return on Investment) for a portfolio of change projects within an organization.

We can, therefore, speak of synergy, which involves the accurate selection of problems to be solved within the organization. This creates an accelerated organizational learning loop, where solutions that are most needed within the organizational process system are systematically sought. Synergy means that none of the perspectives are lost or marginalized: neither the strategic development perspective, nor the technical issues along with production methods and technology bundles, but also, and perhaps above all, the social perspective along with the empowerment of operators. No single method or methodology can sufficiently guarantee this on its own.

References

Basu, R. (2009). *Implementing Six Sigma and Lean*. Taylor & Francis.

Dettmer, W. H. (2001). *Beyond Lean Manufacturing: Combining Lean and the Theory of Constraints for Higher Performance*. Systems International.

Eckes, G. (2016). *Rewolucja Six Sigma: Jak General Electric i inne przedsiębiorstwa zmieniły proces w zyski*. MT Biznes.

Franken, J. C. M., van Dun, D. H., & Wilderom, C. P. M. (2024). Kaizen Event process factors for operational performance improvement: An archival study. *Production Planning & Control*, 36(10), 1–17.

George, M. L. (2002). *Lean Six Sigma: Combining Six Sigma Quality with Lean Production Speed*. New York: McGraw-Hill.

Goldratt, E. M. (1984). *The Goal: A Process of Ongoing Improvement*. North River Press.

Goldratt, E. M. (1990). *Theory of Constraints*. North River Press.

Goldratt, E. M., & Cox, J. F. (2004). *The Goal: A Process of Ongoing Improvement* (3rd ed.). North River Press.

Hamrol, A. (2018). *Strategie i praktyki sprawnego działania: Lean Six Sigma i inne*. Wydawnictwo Naukowe PWN.

Imai, M. (2007). *Kaizen: Klucz do konkurencyjnego sukcesu Japonii*. MT Biznes.

Kose, T., & Agdeniz, S. (2018). Integration of theory of constraints, lean manufacturing and six sigma – The TLS model. In *7th International Congress on Current Debates in Social Sciences*, Istanbul.

McCarty, T., Bremer, M., Daniels, L., & Gupta, P. (2004). *The Six Sigma Black Belt Handbook*. Motorola University.

Nave, D. (2002). How to compare Six Sigma, Lean and the theory of constraints. *Quality Progress*, 35(3), 73–78.

Owieczka, R. (2024). Daily KAIZEN pierwszym krokiem do zmiany kultury organizacji. In M. Lis & M. Bryke (Eds.), *Transformacja KAIZEN w perspektywie przemiany kulturowej organizacji*. Akademia WSB. pp. 37–45.

Pyzdek, T., & Keller, P. A. (2010). *The Six Sigma Handbook: A Complete Guide for Green Belts, Black Belts, and Managers at All Levels* (3rd ed.). McGraw-Hill.

Ronen, B., & Pass, S. (2007). *Focused Operations Management: Achieving More with Existing Resources*. John Wiley & Sons.

Six Sigma: A Complete Step-by-Step Guide: A Complete Training & Reference Guide for White Belts, Yellow Belts, Green Belts, and Black Belts. (2018). Council for Six Sigma Certification.

Van Kollenburg, T. (2021). *Lean Green Belt: Improvement as Practical Skill in the 21st Century*. Learning Lean.

Vo, B., Kongar, E., & Suárez Barraza, M. F. (2019). Kaizen event approach: A case study in the packaging industry. *International Journal of Productivity and Performance Management*, 68(7), 1353–1372.

8

The Pull System: Kanban and DBR

8.1 General Concept

This chapter presents two fundamental methods for managing flow in a production system. The first method is Kanban, derived from Lean Management. It is a visual method of controlling workflow in production systems. The second method is DBR (Drum-Buffer-Rope), derived from the Theory of Constraints and aimed at synchronizing production with constraints.

Both flow control methods implement the same coordination system based on the "pull" principle, that is, production in response to actual demand. Kanban is an operational tool that works at the local level, while DBR is a strategic tool that identifies constraints and synchronizes the entire process around its capabilities.

The reader learns the basic principles of both methods, their key assumptions, and typical areas of application. This chapter discusses how both methods, despite differences in structure, can work together to optimize workflow, increase predictability, and minimize waste. This chapter concludes with a practical example showing the integration of both methods in a functioning production system in the printing industry.

8.2 Introduction

Production management is based, among other things, on value stream management. From the moment a customer places an order to the moment the product reaches them, every action should create value or be eliminated if it does not. This is the fundamental basis of Lean Management, pioneered by Toyota.

Traditional production management is based on the push principle, that is, producing goods in advance based on forecasts. A different method of flow management is the pull system. It is one of the basic principles of organizing smooth flow in Lean. The pull system means that production is based

DOI: 10.1201/9781003655237-8

on actual customer demand, thus "pulling" the product from the last stage of the process, freeing up space, and giving the signal to start production. Production organization in accordance with the pull system eliminates overproduction, which in Lean is considered one of the worst types of waste.

Although the pull system seems simple and elegant, its practical implementation can take various forms. These are Kanban and DBR, which control flow in different ways.

8.3 Comparison of Kanban and DBR Systems

Kanban is a tool derived from the Lean Management methodology. In Japanese, Kanban means card or board. It is primarily used for visual workflow management in relation to actual demand. The task of Kanban is to implement "Just-in-Time" production. Its primary goal is to reduce overproduction, which is one of the key types of waste defined in Lean Management, as well as to maintain a stable flow of value and shorten lead time.

Kanban was developed within the Toyota Production System. Its purpose was to regulate the flow of parts between successive processes. Kanban cards were introduced, which operate on the pull system principle. Tasks are transferred to the next stage when that stage has completed its work and space has become available within its work in process limit. In other words, production is "pulled" from the previous stage. The pull system ensures that production takes place in line with actual demand, prevents the accumulation of excessive inventory between stages, and synchronizes the pace of processes.

As already mentioned, Kanban takes the form of various cards, each of which performs a specific function in controlling the production flow. In a production environment, the basic type of card is Production Kanban, which signals production needs between processes in the value stream. When a production process receives such a card, it is a signal to start manufacturing. The next one is Withdrawal Kanban, which is a signal to collect materials from the previous process. In a production environment, both types of cards work closely together. In more advanced production systems, there are also Signal Kanban, Supplier Kanban, Emergency Kanban, and others (The Productivity Press Development Team, 2009).

The flow of value in Kanban should be visualized on a board (physical or digital) showing the successive stages of the process. Kanban cards move between stages as work progresses. The key principle is to define work in progress (WIP), that is, the number of tasks that can be between processes. Kanban aims to maintain a smooth, stable, and above all, predictable flow. Key performance indicators for Kanban include lead time, cycle time, and throughput.

Kanban originated in the manufacturing environment, but its principles are so universal that they have been used in knowledge work, where the object of sale is not a physical product, but information, software, projects, and other items. Kanban is a tool often used in logistics, IT (Kanban software), and administration. Kanban is, in a sense, a self-directed horizontal flow. Employees control the pace of task completion themselves, without the need for constant intervention from a manager.

Another method of controlling workflow, this time derived from the Theory of Constraints, is DBR, developed by Eliyahu M. Goldratt in the 1980s. Its main purpose is to synchronize the entire system with its critical point—the constraint—minimizing inventory and WIP. Kanban limits WIP by indicating the maximum WIP between processes. DBR, on the other hand, sets the pace of the system, protecting the bottleneck with a buffer and controlling the start of other processes with a rope (Blackstone, 2010, p. 149).

In DBR, the bottleneck determines the pace of work for the other resources before and after it. This bottleneck affects the throughput of the entire production system and determines the size and frequency of material deliveries. Therefore, all planning activities should focus on maximizing production flows by subordinating other resources to the bottleneck (Dettmer, 2001, p. 26).

The DBR method consists of three elements. The first is the Drum, which sets the pace for the entire system according to the bottleneck's rate of operation. The constraint determines the capacity of the entire system. It can be a machine on a production line, a workstation with the highest demand relative to availability, or market demand (more on other types of constraints in Chapter 5).

The second—Buffer—ensures stable flow and continuity of production at the bottleneck (constraint buffer). It is placed before the constraint to protect it from upstream disruptions. In this way, the bottleneck is ensured maximum utilization. It most often occurs as a buffer time (material is delivered in advance of the bottleneck) or a buffer stock (WIP located before the bottleneck). The buffer size is set as a minimum and depends on the variability of the process. It ranges from 30% to 50% of the task's production time. The buffer is controlled and monitored. However, this is not wasteful, but rather a safeguard for the throughput of the entire production system. DBR also distinguishes a shipping buffer, which protects the customer and secures the delivery date against possible disruptions occurring, among other things, in the production process. The buffer consumption level is monitored using color coding. Green means that the buffer level is stable. Yellow means that there are some disruptions in the upstream. Red means that the bottleneck is at risk of losing throughput.

The last element of DBR is Rope. It sets the rhythm of delivering materials or semi-finished products to processes upstream of the bottleneck in order to eliminate overproduction on non-constraint resources. This rhythm is consistent with the actual pace of the constraint. It acts like a physical rope that connects the constraint to the beginning of the process and signals that new material can be admitted when the constraint is ready.

DBR is consistent with the logic of the Five Focusing Steps method: Step 1—identification of the constraint, that is, the drum, which sets the pace of production; Step 2—exploitation of the constraint, that is, developing buffer levels to effectively utilize the bottleneck; Step 3—using a tool called rope to subordinate the schedule for releasing materials to the process in order to maintain a constant buffer capacity (Blackstone, 2010, p. 149).

DBR as a workflow control method works well in processes that are variable but repeatable, where work is carried out in a long sequence, that is, the product must go through many consecutive steps before it is fully completed, and the order completion time and lead time are known.

8.4 Synergy between Kanban and DBR to Maximize Flow

How can you ensure that your production system produces exactly as much as actual demand dictates? As described in the previous section, there are two ways to do this. The first possible method is the Kanban approach, which aims to ensure that workstations do not produce more than the next one can accept. The second possible method is DBR, whose task is to identify the bottleneck in the entire process that most constrains everything else and dictates the rhythm of the entire system. Both methods actually lead to the same goal, which is to produce exactly what the customer needs, rather than "just in case."

These two flow control methods are based on the same fundamental Lean principle, which was formulated by Womack and Jones (1996) as a pull-based coordination system (more details in Chapter 2). This principle states that you should produce and deliver exactly what the customer wants and when they want it. You should not produce "mindlessly" just because the machines have to run. Production should be based on actual demand.

Kanban is characterized by a distributed pull system. This means that each workstation makes decisions based on what it sees in front of and behind it. DBR, on the other hand, is characterized by a centralized pull system. This means that there is one reference point in the system that sets the pace of work, and the rest of the system must adapt to it. The difference between Kanban and DBR is not that one is better or worse than the other. The difference is that each operates at a different level.

A manager observing the Kanban board on the production floor sees the workflow at the operational level. They know exactly what the operator is doing or should be doing. Now, if the same manager looked at the same processes from "a bird's eye view," they would be able to estimate the number of orders that can be produced and what delivery time can be promised to the customer. In this way, the manager sees the workflow at the strategic level.

Therefore, Kanban and DBR are not mutually exclusive. The former operates at the so-called local level, and the latter at the so-called global level. In other words, DBR sets the pace (rhythm) for the entire system, and each position in that system must know what to do at that pace. Together, they lead to the elimination of overproduction and chaos and a stable workflow.

The strength of these methods stems from the complementarity of their perspectives. Each method operates at a different level, and combining them provides a complete workflow management mechanism. The strength of these methods also lies in the fact that DBR sets the parameters for the entire system (throughput), while Kanban controls the flows between the individual stages of the system.

The combination of both methods also affects variability control. DBR, through a buffer (usually buffer time or buffer stock), strategically protects against any disruptions resulting from system malfunctions (machine failures, employee absences, and other unplanned interruptions). Kanban protects other processes by introducing control cards and WIP limits. It can be said that flow protection occurs at both the global and local levels. The main benefit of volatility regulation is a reduction in the risk of downtime and sudden spikes in workload.

The integration of Kanban and DBR affects production planning and scheduling. The rope in DBR indicates how much work can be allowed into the system (knowing the constraint limit), while Kanban regulates the pace of individual processes. DBR controls work at the system level and Kanban at the local process level. Working, they create the ability to control flow while eliminating bottlenecks and maintaining stable flow. Neither of these methods would achieve this on its own.

In summary, both methods make the production system transparent and resistant to disruptions. Generally speaking, DBR identifies the most important constraint in the system and indicates how to protect it. Kanban, on the other hand, shows how to organize the rest of the process so that the flow is smooth and protects against excessive inventory. Together, they reduce waste, maintain a stable flow, shorten lead times, and protect the main constraint that determines throughput.

8.5 Kanban and DBR—A Business Case Study

The object of the study is a company operating in the printing industry. The company manufactures cardboard packaging for the food and medical industries. The production system under study consists of four basic processes, such as (1) offset printing, (2) die stamping, (3) die cutting, and (4) folding and gluing the product. The production system is characterized by a highly dependent workflow, in which each process is dependent on the previous one. Disruptions occurring in one process translate into downstream problems. The aim of this study is to compare the Kanban and DBR methods using an empirical case study.

In the production system under study, it was noted that the bottleneck was the folding and gluing process. This process was characterized by the lowest hourly productivity, and high levels of WIP were also created. In the company under study, the start of production was mainly based on daily plans and other often subjective decisions made by operational employees.

The first stage was the implementation of the DBR method, which was designed to control the workflow globally. The first element—Drum, that is, the rhythm of the entire system, which is consistent with the rhythm of the bottleneck—was determined based on the actual throughput of the bottleneck. The bottleneck operates at a rate of 18,000 product units per shift. The determined value determines the throughput of the entire system. Upstream orders could not be fulfilled if they were not included in the constraint schedule.

Next, a buffer was established to ensure stable flow and continuity of production at the bottleneck. It was placed as a physical queue of cutouts after the die cutting process. The buffer stock was 2 pallets per order. This is a protective level of restriction designed to counteract upstream disruptions and minimize excessive priority changes.

Another element is Rope, which sets the rhythm for delivering materials or semi-finished products to processes preceding folding and gluing in order to eliminate overproduction on resources that are not a constraint. Rope connects the constraint with the beginning of the process, that is, offset printing, and signals that new material can be admitted when the constraint is ready. Rope is designed to limit the start of production in upstream processes. It is a clear, global production flow cycle. In line with the assumptions of the DBR method, at this point the pace of the entire production system has been stabilized and lower WIP limits are noticeable.

In the next step, the Kanban method was implemented at the local variability level. The purpose of Kanban is to improve cooperation between processes occurring before the bottleneck, that is, offset printing, die stamping, die cutting, as well as to standardize production batches. At this stage, two types of Kanban cards were introduced. The first is Production Kanban, which signals the need to start a new batch at a given stage. The second is Withdrawal Kanban, which signals the need to collect material for the next stage.

Next, WIP limits were calculated. It was determined that after the offset printing process, WIP is 3 pallets, after the die stamping process—2 pallets, and according to Buffer DBR after the die cutting process—2 pallets. The limits estimated in this way stabilized the flow throughout the entire production system. Both the DBR and Kanban methods allowed for the development of a pull-based work system. Processes upstream of the bottleneck

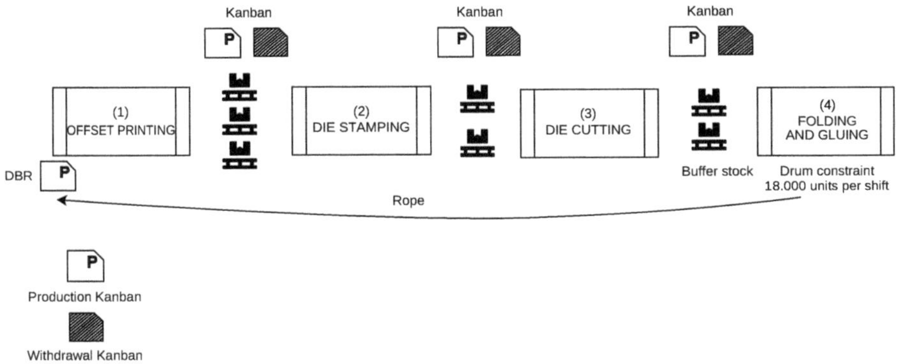

FIGURE 8.1
Optional utilization of Kanban or DBR.

could only produce when they received a Kanban card from the constraint. As a result, the offset printing process did not start production at any time, the die stamping process did not generate excessive inventory, and the die cutting process replenished the DBR buffer on an ongoing basis. Figure 8.1 shows the optional use of Kanban and DBR in the production system under study.

Both the Kanban and DBR methods enable the system to function on two levels. The first is the global level. Thanks to DBR, the production process runs at a pace determined by the bottleneck. The second is the local level. Thanks to the Kanban method, the flow between processes that do not constitute a bottleneck has been regulated, and materials are delivered evenly.

In practice, this means that the flow can proceed as an upper stream, that is, using the Kanban method, where it will be controlled by demand signals between processes. The second solution is that the flow can proceed as a lower stream, that is, according to DBR principles, where the pace is dictated by the Drum, and the Rope controls the work that is allowed into the system.

However, due to the specific nature of printing, whose production system is characterized by medium print runs and high order variability, managers believe that DBR is a more advantageous tool at the system level. The DBR method better stabilizes the flow, which focuses on a single constraint (folding and gluing).

References

Blackstone, J. H. (2010). A review of literature on Drum-Buffer-Rope, buffer management and distribution. In J. F. III Cox & J. G. Schleier (Eds.), *Theory of Constraints Handbook*. McGraw-Hill. pp. 145–174.

Dettmer, W. H. (2001). *Beyond Lean Manufacturing: Combining Lean and the Theory of Constraints for Higher Performance*. Systems International. Port Angeles – USA.
The Productivity Press Development Team. (2009). Kanban na hali produkcyjnej. ProdPress.com.
Womack, J. P., & Jones, D. T. (1996, September – October). *Lean Thinking: Banish Waste and Create Wealth in Your Corporation*. Harvard Business Review.

Index

For Product Safety Concerns and Information please contact our EU
representative GPSR@taylorandfrancis.com
Taylor & Francis Verlag GmbH, Kaufingerstraße 24, 80331 München, Germany

www.ingramcontent.com/pod-product-compliance
Ingram Content Group UK Ltd.
Pitfield, Milton Keynes, MK11 3LW, UK
UKHW020207200926
475702UK00004B/657
9781041104100